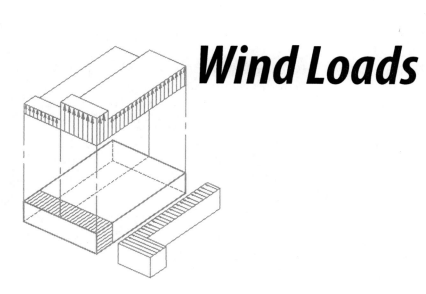

Wind Loads

Other Titles of Interest

Minimum Design Loads for Buildings and Other Structures, ASCE/SEI 7-05. (ASCE Standard, 2006). Provides requirements for general structural design and includes means for determining various loads and their combinations. Includes commentary. (ISBN 978-0-7844-0831-5)

***Snow Loads: Guide to the Snow Load Provisions of ASCE 7-05,* by Michael O'Rourke.** (ASCE Press, 2007). Presents a detailed, authoritative interpretation of the snow load provisions of ASCE/SEI 7-05, including worked examples and FAQs. (ISBN 978-0-7844-0857-5)

***Seismic Loads: Guide to the Seismic Load Provisions of ASCE 7-05,* by Finley A. Charney.** (ASCE Press, 2010). Offers an authoritative interpretation of the seismic load provisions of ASCE/SEI 7-05, including worked examples and FAQs. (ISBN 978-0-7844-1076-9)

Also:

***Wind Tunnel Studies of Buildings and Structures,* edited by Nicholas Isyumov.** (ASCE Manual No. 67, 1999). Assists architects and engineers to improve the reliability of structural performance and to achieve cost effectiveness. (ISBN 978-0-7844-0319-8)

***In the Wake of Tacoma: Suspension Bridges and the Quest for Aerodynamic Stability,* by Richard Scott** (ASCE Press, 2001). Surveys changes in the design of suspension bridges evolving from the 1940 collapse of the first Tacoma Narrows Bridge. (ISBN 978-0-7844-0542-0)

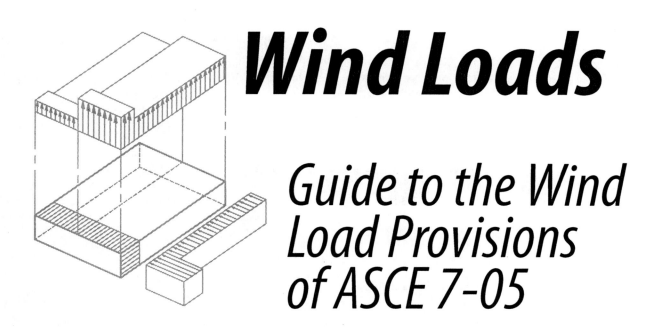

Wind Loads

Guide to the Wind Load Provisions of ASCE 7-05

Kishor C. Mehta, Ph.D., P.E.
William L. Coulbourne, P.E.

Library of Congress Cataloging-in-Publication Data

Mehta, Kishor C.
  Wind loads : guide to the wind load provisions of ASCE 7-05 / Kishor C. Mehta, William L. Coulbourne.
     p. cm.
  Includes bibliographical references and index.
  ISBN 978-0-7844-0858-2
     1. Wind-pressure.   2. Buildings–Standards–United   States.   3. Buildings–Aerodynamics. 4. Gust loads. 5. Structural engineering. I. Coulbourne, William L. II. Title.
  TH891.M453 2010
  690'.1—dc22

                                        2010009410

Published by American Society of Civil Engineers
1801 Alexander Bell Drive
Reston, Virginia 20191
www.pubs.asce.org

Any statements expressed in these materials are those of the individual authors and do not necessarily represent the views of ASCE, which takes no responsibility for any statement made herein. No reference made in this publication to any specific method, product, process, or service constitutes or implies an endorsement, recommendation, or warranty thereof by ASCE. The materials are for general information only and do not represent a standard of ASCE, nor are they intended as a reference in purchase specifications, contracts, regulations, statutes, or any other legal document.

ASCE makes no representation or warranty of any kind, whether express or implied, concerning the accuracy, completeness, suitability, or utility of any information, apparatus, product, or process discussed in this publication, and assumes no liability therefor. This information should not be used without first securing competent advice with respect to its suitability for any general or specific application. Anyone utilizing this information assumes all liability arising from such use, including but not limited to infringement of any patent or patents.

ASCE and American Society of Civil Engineers—Registered in U.S. Patent and Trademark Office.

Photocopies and reprints. You can obtain instant permission to photocopy ASCE publications by using ASCE's online permission service (http://pubs.asce.org/permissions/requests/). Requests for 100 copies or more should be submitted to the Reprints Department, Publications Division, ASCE (address above); e-mail: permissions@asce.org. A reprint order form can be found at http://pubs.asce.org/support/reprints/.

Copyright © 2010 by the American Society of Civil Engineers.
All Rights Reserved.

ISBN 978-0-7844-0858-2

Manufactured in the United States of America.
18 17 16 15 14 13 12 11 10   1 2 3 4 5

Contents

Preface vii
Table of Conversion Factors viii

Chapter 1 Introduction 1
        Objective of the Guide 2
        Significant Changes and Additions 2
        Limitations of Standard 3
        Technical Literature 5

Chapter 2 Wind Load Provisions 7
        Format 7
        Design Procedures 8
        Method 3, Wind Tunnel Procedure 12
        Equations for Graphs 13

Chapter 3 Examples 23
        Example 1: 30-ft × 60-ft × 15-ft Commercial Building
          with Concrete Masonry Unit Walls 23
        Example 2: Ex. 1 Using Simplified Procedure 31
        Example 3: 100-ft × 200-ft × 160-ft High Office Building 35
        Example 4: Office Building from Ex. 3 Located
          on an Escarpment 47
        Example 5: 2,500-ft^2 House with Gable/Hip Roof 50
        Example 6: 200-ft × 250-ft Gable Roof Commercial/
          Warehouse Building Using Buildings of All Height
          Provisions 59
        Example 7: Building from Ex. 6 Using Low-Rise Building
          Provisions 72
        Example 8: 40-ft × 80-ft Commercial Building
          with Monoslope Roof with Overhang 80
        Example 9: U-Shaped Apartment Building 93
        Example 10: 50-ft × 20-ft Billboard Sign
          on Poles (Flexible) 60 ft Above Ground 104
        Example 11: Domed Roof Building 111
        Example 12: Unusually Shaped Building 119
        Example 13: 30-ft × 60-ft Open Building with Gable Roof ... 130

Chapter 4 Frequently Asked Questions 137

References 147
Index 153
About the Authors 159

Preface

This guide is designed to assist professionals in the use of the wind load provisions of ASCE/SEI Standard 7-05, *Minimum Design Loads for Buildings and Other Structures*, published by the American Society of Civil Engineers (ASCE). The guide is a revision of the *Guide to the Use of Wind Load Provisions of ASCE 7-02*, reflecting the significant changes made to wind load provisions when the previous version of the Standard, SEI/ASCE 7-02, was updated. The guide contains 13 example problems worked out in detail, which can provide direction to practicing professionals in assessing wind loads on a variety of buildings and other structures. Every effort has been made to make these illustrative example problems correct and accurate. The authors would welcome comments regarding inaccuracies, errors, or different interpretations. The views expressed and interpretation of the wind load provisions made in the guide are those of the authors and not of the ASCE 7 Standards Committee or of the American Society of Civil Engineers.

Table of Conversion Factors

U.S. customary units	International System of Units (SI)
1 inch (in.)	25.4 millimeters (mm)
1 foot (ft)	0.3048 meter (m)
1 statute mile	1.6093 kilometers (km)
1 square foot (ft^2)	0.0929 square meter (m^2)
1 cubic foot (ft^3)	0.0283 cubic meter (m^3)
1 pound (lb)	0.4536 kilogram (kg)
1 pound (force)	4.4482 newtons (N)
1 pound per square foot (lb/ft^2)	0.0479 kilonewton per square meter (kN/m^2)
1 pound per cubic foot (lb/ft^3)	16.0185 kilograms per cubic meter (kg/m^3)
1 degree Fahrenheit (°F)	1.8 degrees Celsius (°C)
1 British thermal unit (Btu)	1.0551 kilojoules (kJ)
1 degree Fahrenheit per British thermal unit (°F/Btu)	1.7061 degrees Celsius per kilojoule (°C/kJ)

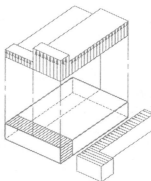

Chapter 1
Introduction

The American Society of Civil Engineers (ASCE) publication, *Minimum Design Loads for Buildings and Other Structures*, ASCE/SEI Standard 7-05, is a consensus standard. It originated in 1972 when the American National Standards Institute (ANSI) published a standard with the same title (ANSI A58.1-1972). That 1972 standard was revised ten years later, containing an innovative approach to wind loads for components and cladding (C&C) of buildings (ANSI A58.1-1982). Wind load criteria were based on the understanding of aerodynamics of wind pressures in building corners, eaves, and ridge areas, as well as the effects on pressures of area averaging.

In the mid-1980s, the ASCE assumed responsibility for the Minimum Design Loads for Buildings and Other Structures Standards Committee, which establishes design loads. The document published by ASCE (ASCE 7-88) contained design load criteria for live loads, snow loads, wind loads, earthquake loads, and other environmental loads, as well as load combinations. The ASCE 7 Standards Committee has voting membership of close to 100 individuals representing all aspects of the building construction industry. The criteria for each of the environmental loads are developed by respective task committees.

The wind load criteria of ASCE 7-88 (ASCE, 1990) were essentially the same as ANSI A58.1-1982. In 1995, ASCE published ASCE 7-95. This version contained major changes in wind load criteria: the basic wind speed averaging time was changed from fastest-mile to 3-second gust. This in turn necessitated significant changes in boundary-layer profile parameters, gust effect factor, and some pressure coefficients. A *Guide to the Use of the Wind Load Provisions of ASCE 7-95* (Mehta and Marshall, 1998) was published by ASCE to assist practicing professionals in the use of wind load criteria of ASCE 7-95.

In 2000, ASCE published a revision of ASCE 7-95 with updated wind load provisions. The document is termed ASCE 7-98 and has the same title (ASCE, 2000). The International Building Code (ICC, 2000) adopted the wind load criteria of ASCE 7-98 by reference. This was a major milestone since it had the potential to establish a single wind load criterion for design of all buildings and structures for the entire United States. A *Guide to the Use of the Wind Load Provisions of ASCE 7-98* (Mehta and Perry, 2000) was published soon after publication of ASCE 7-98. This document was updated as *Guide to the Use of the Wind Load Provisions of ASCE 7-02* (Mehta and Delahay, 2004) after publication of the revised standard ASCE 7-02.

In 2005, the new standard, ASCE 7-05, was published. This guide is designed to assist practicing professionals in the use of wind load criteria of ASCE 7-05.

1.1 Objective of the Guide

The objective of this guide is to provide direction in the use of wind load provisions of ASCE 7-05 (referred to as "the Standard"). The Commentary of ASCE 7-05 (Chapter C6) contains a good background and discussion of the wind load criteria; that information is not repeated in this document. Rather, this guide contains two important items to assist the users of ASCE 7-05: (1) examples, and (2) frequently asked questions.

The guide contains 13 worked examples. Sufficient details of calculation of wind loads are provided to help the reader properly interpret the wind load provisions of the Standard. Chapter 6 of the Standard, as well as the figures and tables of the Standard, are cited liberally in the examples. **It is necessary to have a copy of ASCE 7-05 to follow the examples and work with this guide.** A copy of ASCE 7-05 can be ordered by calling 1-800-548-ASCE or on the Internet at *www.pubs.asce.org*.

1.2 Significant Changes and Additions

The wind load provisions of Chapter 6.0 were revised in ASCE 7-05 using recent research and development achievements. The major additions involve roof pressures over open buildings, loads on rooftop equipment, and expansion of the commentary.

The basic approach to assessing wind loading has not been changed. Significant changes/additions can assist the design process and are listed below.

- Simplified procedure for MWFRS is permitted only if torsion loading consideration is exempted or does not control.

- Simplified procedure is permitted for hilly terrain if topographic effects are taken into account.
- Upwind distances are changed for Exposures B and D.
- In wind-borne debris regions, impact resistant glazing or protective covering in the bottom 60 ft of buildings is made mandatory.
- Parapet pressure coefficients for MWFRS loading are revised.
- For open buildings, pressure coefficients for monoslope, pitched and troughed roofs are added.
- Force coefficients for freestanding solid walls and signs are added.
- Rooftop equipment wind load provisions for buildings less than 60 ft are added.
- Commentary Chapter C6 for wind loads is expanded to provide guidance in hurricane wind speeds, torsional sensitivity of buildings, assessment of fundamental frequency, and other items.

As noted above, the basic methodology of the Standard remains the same as in ASCE 7-02. Additional information on the changes can be found in the Commentary of the Standard and from references.

1.3 Limitations of Standard

The possible shortcomings or limitations of the Standard are directly dependent on accurate knowledge of parameters and factors used in the algorithms that define the wind loads for design applications. Limitations of some of the significant parameters are given below.

1.3.1 Assessment of Wind Climate

The current Standard provides a more realistic description of wind speeds than did the previous editions of the 1970s and 1980s. Perhaps the most serious limitations are that design speeds are not referenced to direction, and potential wind speed anomalies are defined only in terms of special wind regions. These special wind regions include mountain ranges, gorges, or river valleys. Unusual winds may be encountered in these regions because of orographic effects or because of the channeling of wind. The Standard permits climatological studies using regional climatic data and consultation with a wind engineer and/or a meteorologist.

Tornado winds are not included in development of the basic wind speed map (Figure 6-1 of the Standard) because of their relatively rare occurrence at a given location. Intense tornadoes can have ground level wind speeds in the range of 150 to 200 mph; however, the annual probability of exceedance of this range of wind speeds may be less than 1×10^{-5} (mean recurrence interval exceeding 100,000 years). Special structures and storm shelters can be designed to resist tornado winds if required.

1.3.2 Limitations in Evaluating Structural Response

Given that the majority of buildings and other structures can be treated as rigid structures, the gust effect factor specified in the Standard is adequate. For dynamically sensitive buildings and other structures, a gust effect factor, G_f, is given. The formulation of gust effect factor, G_f, is primarily for buildings; it is not always applicable to other structures. It should be noted that the gust effect factor, G_f, is based on along-wind buffeting response.

Vortex shedding is almost always present with bluff-shaped cylindrical bodies. It can become a problem when the frequency of shedding is close to, or equal to, the frequency of the first or second transverse modes of the structure. The intensity of excitation increases with aspect ratio (height-to-width or length-to-breadth) and decreases with increasing structural damping. Structures with low damping and with an aspect ratio of 8 or more may be prone to damaging vortex excitation. If across-wind or torsional excitation appears to be a possibility, expert advice should be obtained.

Another limitation with respect to evaluating structural response is that the Standard does not define acceptable design wind speeds for serviceability states (e.g., deflection, dynamic sway). Table C6-7 in the Commentary provides conversion factors for determining appropriate wind speeds for mean recurrence intervals of 5 to 500 yr.

1.3.3 Limitations in Shapes of Buildings and Other Structures

The pressure and force coefficients given in the Standard are limited. Many of the structural shapes (e.g., "Y," "T," and "L" shapes) or buildings with stepped elevations are not included (except as shown in Figure 6-12). Fortunately, this information may be found in other sources (see **Table G1–1** of this guide).

When coefficients for a specific shape are not given in the Standard, the designer is encouraged to use values that are available in the literature. However, the use of prudent judgment is advised, and the following caveats must be addressed:

1. Were the coefficients obtained from proper turbulent boundary layer wind tunnel (BLWT) tests, or were they generated under conditions of relatively smooth flow?
2. The averaging time used must necessarily be considered in order to determine whether the coefficients are directly applicable to the evaluation of design loads or whether they need to be modified.
3. The reference wind speed (fastest-mile, hourly mean, 10-min mean, 3-s gust, etc.) and exposure category under which the data are generated must be established in order to properly compute the velocity pressure, q.
4. If an envelope approach is used, the coefficients should be appropriate for all wind directions. If, however, a directional approach is indicated, then the applicability of the coefficients as a function of wind direction needs to be ascertained. A major limitation in the

use of directional coefficients is that their adequacy for other than normal wind directions may not have been verified.

1.4 Technical Literature

There has been a vast amount of literature published on wind engineering during the past three decades. Most of it is in the form of research papers in the *Journal of Wind Engineering and Industrial Aerodynamics, Journal of Structural Engineering, Proceedings of the International Conferences on Wind Engineering* (a total of twelve), *Proceedings of the Americas Conferences on Wind Engineering* (a total of eleven), *Proceedings of the Asia-Pacific Conferences on Wind Engineering* (a total of six) and *Proceedings of the European-African Conferences on Wind Engineering* (five). The literature is extensive and scholarly; however, it is not always in a format that can be used by practicing professionals.

Several textbooks, handbooks, standards and codes, reports, and papers contain material that can be used to determine wind loads. Selected items are identified in **Table G1–1** of this guide. The items are listed by subject matter for easy identification. Detailed references for these items are given in the citations in References, of this guide.

Table G1-1 Technical Literature

Subject	Selected reference material (see References section of this guide)
Wind effects on buildings and structures	Newberry and Eaton (1974) Lawson, vols. 1 and 2 (1980) Cook, parts 1 and 2 (1985) Holmes, Melbourne, and Walker (1990) Liu (1991) Simiu and Scanlan (1996) Holmes (2001)
Foreign codes and standards	NRCC (2005) British Standard BS 6399 (1997) Eurocode 1 (1998) ISO (1997) Australian/New Zealand Standard AS/NZS 1170.2 (2002)
Wind tunnel testing	Reinhold (1982) ASCE (1999)
General wind research	ASCE (1961) Cermak (1977) Davenport, Surry, and Stathopoulos (1977, 1978) Simiu (1981)
Pressure and force coefficients	ASCE (1961, 1997) Hoerner (1965)
Tornadoes, shelter design	FEMA TR83-A (1980) Minor (1982) Minor, McDonald, and Mehta (1993) McDonald (1983) FEMA 320 (2008) FEMA 361 (2008)
Impact resistance protocol	SBCCI (1999) ASTM E1886-05 (2005) ASTM E1996-08e2 (2008) Miami/Dade County Building Code Compliance Office Protocol PA 201-94 and PA 203-94

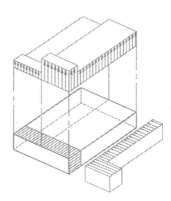

Chapter 2
Wind Load Provisions

2.1 Format

The designer is given three options for evaluating the design wind loads for buildings and other structures:

Method 1   Simplified Procedure, as specified in Section 6.4 of the Standard, for buildings meeting certain specific requirements. The requirements are set for main wind force-resisting system (MWFRS) and components and cladding (C&C), respectively.

Method 2   Analytical Procedure, of Section 6.5 of the Standard, applicable to buildings and other structures.

Method 3   Wind Tunnel Procedure, which meets certain test conditions as specified in Section 6.6 of the Standard.

The simplified and analytical procedures (see Sections 6.4.2 and 6.5.3, respectively) provide specific steps to be followed in the determination of wind loads on MWFRS and C&C separately. MWFRS is defined in Section 6.2 as an assemblage of structural elements assigned to provide support and stability for the overall structure; it always receives wind loading from more than one surface. Components and cladding receive wind loads directly and generally transfer the load to other components or to the MWFRS.

Equations for the determination of wind loads using the analytical procedures are given in the body of the text of the Standard.

Equations for the graphs of Figures 6-11A through 6-17 in the Standard are given in Section 2.4 of this guide because interpolation using these graphs, as presented in the Standard, is difficult.

2.2 Design Procedures

2.2.1 Velocity Pressure

The first step in using Method 2, Analytical Procedure, is to determine the appropriate parameters for evaluating the velocity pressure, q.

Velocity pressure, q, at any height above ground and at mean roof height is obtained by the following equation:

$$q_z = 0.00256 K_z K_{zt} K_d V^2 I \text{ (lb/ft}^2) \qquad \text{(Eq. 6-15)}$$

where

q = Effective velocity pressure to be used in the appropriate equations to evaluate wind pressures for MWFRS and C&C; q_z at any height, z, above ground; q_h is based on K_h at mean roof height, h

K_z = Exposure velocity pressure coefficient, which reflects change in wind speed with height and terrain roughness (see Section 6.5.6 and Table 6-3 of the Standard)

K_{zt} = Topographic factor, which accounts for wind speed-up over hills and escarpments (see Section 6.5.7 and Figure 6-4 of the Standard)

K_d = Directionality factor (see Section 6.5.4.4 and Table 6-4 of the Standard)

V = Basic wind speed, which is the 3-s gust speed at 33 ft above ground for Exposure Category C and is associated with an annual probability of 0.02 (50-yr mean recurrence interval) (see Section 6.5.4 and Figure 6-1 of the Standard)

I = Importance factor, which adjusts wind speed associated with annual probability of 0.02 (50-yr mean recurrence interval) to other probabilities (25- or 100-yr MRI) (see Section 6.5.5 and Table 6-1 of the Standard)

2.2.2 Method 1, Simplified Procedure

Method 1 was introduced in ASCE 7-98 for simplifying evaluation of design loads for common regular-shaped buildings. Since then, provisions of this method have been revised significantly. The restrictions for using the simplified procedure are set for MWFRS and C&C in Sections 6.4.1.1 and 6.4.1.2, respectively.

Tabulated wind pressure values are provided in Figure 6-2 for MWFRS and Figure 6-3 for C&C. For MWFRS, Method 1 combines the windward and leeward pressures into a net horizontal wind pressure on the walls (internal pressures cancel). The maximum uplift on low-slope roofs for MWFRS is based on a positive internal pressure as the controlling case and is applied on horizontal projection of the roof surface. For C&C, values are provided only for enclosed buildings and represent the net pressure (sum of external and internal pressures) applied normally to surfaces. The following values have been assumed in the preparation of the tabulated values:

$h = 30$ ft

Exposure B, $K_z = 0.70$

$K_d = 0.85$

$G = 0.85$

$K_{zt} = 1.0$

$I = 1.0$

$GC_{pi} = \pm 0.18$ (enclosed building)

MWFRS pressure coefficients from Figure 6-10

C&C pressure coefficients from Figure 6-11A–D

Multiplying factor λ is given for different mean roof heights and exposure classifications in Figures 6-2 and 6-3 of the Standard. For importance factors other than $I = 1.0$, tabulated pressure values should be multiplied by the appropriate value of I. Topographic factor, K_{zt}, if needed, should be determined from Figure 6-4.

2.2.3 Method 2, Analytical Procedure

The analytical procedure for this method is applicable to

1. Buildings of all heights
2. Alternate low-rise buildings with mean roof height less than or equal to 60 ft and as defined in Section 6.2 of the Standard
3. Open buildings and other structures.

The design procedure for each building type is delineated in Section 6.5.3 of the Standard. Velocity pressures, q_z or q_h, are determined in each case using Eq. 6-15 (see Section 6.5.10).

Design pressures for MWFRS and for C&C are determined separately. Generally, C&C design pressures will be higher because of localized high pressures acting over small areas. MWFRS receive wind pressures from several surfaces; hence, with spatial averaging and correlation, the pressures are likely to be smaller than those for C&C.

Calculation of design pressures requires selection of appropriate gust effect factors and pressure or force coefficients. The equation for the evaluation of wind loads guides the user in the selection of appropriate factors and coefficients. Various gust effect factors and pressure and force coefficients specified in the Standard are as follows:

- G Gust effect factor for MWFRS of rigid buildings (all heights) and for other structures (Section 6.5.8.1)
- G_f Gust effect factor for MWFRS of flexible buildings and dynamically sensitive other structures obtained using the procedure given in Section 6.5.8.2 or using a rational analysis (see Section 6.5.8.3)
- C_p External pressure coefficients for MWFRS of closed buildings: all heights (Figure 6-6); domed roof (Figure 6-7); and arched roof (Figure 6-8)
- C_N Net pressure coefficients for MWFRS of open buildings (Figure 6-18A–D) and for C&C of open buildings (Figure 6-19A–C)
- C_f Force coefficients for other structures (Figures 6-20 through 6-23)
- (GC_{pf}) External pressure coefficients for MWFRS of low-rise buildings (Figure 6-10)
- (GC_p) External pressure coefficients for C&C of buildings (Figures 6-11 through 6-17)
- (GC_{pi}) Internal pressure coefficients for MWFRS and C&C of buildings (Figure 6-5)

Sign convention in the Standard is as follows:

+ (plus sign) means pressure acting toward the surface

− (minus sign) means pressure acting away from the surface.

Whenever the sign of "±" is specified, both positive and negative values should be used to obtain design loads. Values of external and internal pressures are to be combined algebraically to obtain the most critical load.

2.2.3.1 Design Pressures for Closed Buildings

Design wind pressures for the MWFRS of rigid buildings of all (any) heights are determined using the following equation:

$$p = qGC_p - q_i(GC_{pi}) \text{ (lb/ft}^2\text{)} \tag{Eq. 6-17}$$

The terms in Eq. 6-17 are defined above. The effective velocity pressure related to internal pressure, q_i, is generally used as q_h (see Section 6.5.12.2.1 of the Standard). Only for high-rise building may it be advantageous to use q_i as defined in Section 6.5.12.2.1 related to positive internal pressure. Use of this term is illustrated in Ex. 3 (Section 3.3 of this guide).

Alternatively, design pressures for MWFRS of low-rise buildings can be determined using the following equation:

$$p = q_h[(GC_{pf}) - (GC_{pi})] \text{ (lb/ft}^2\text{)} \tag{Eq. 6-18}$$

The terms in Eq. 6-18 are defined above. A low-rise building is defined in Section 6.2 of the Standard as a building with mean roof height $h \leq 60$ ft and with mean roof height not exceeding the least horizontal dimension. The design pressures are applied for transverse and longitudinal directions as shown in Figure 6-10. This alternate procedure is appropriate for gable and rectangular buildings, though use of it for any building is permitted. Use of this procedure is illustrated in Ex. 7 (Section 3.7 of this guide).

Design wind pressures for the MWFRS of flexible buildings shall be determined from the following equation:

$$p = qG_fC_p - q_i(GC_{pi}) \text{ (lb/ft}^2\text{)} \tag{Eq. 6-19}$$

where the terms are as defined above, and G_f = gust effect factor as defined in Section 6.5.8.2 of the Standard. The procedure is the same as that for rigid buildings except for determination of gust effect factor, G_f. A flexible building (or structure) is defined in Section 6.2 as one that has fundamental natural frequency less than 1 Hz (period of vibration greater than 1 s). Flexible buildings or structures are affected by the gustiness of the wind and have potential of resonance response. This response results in a large gust effect factor. Calculation of gust effect factor, G_f, for a flexible structure using Eq. 6-8 of Section 6.5.8.2 is illustrated in Ex. 10 (Section 3.10 of this guide).

Design wind pressures on C&C elements of buildings with $h \leq 60$ ft are determined from the following equation:

$$p = q_h[(GC_p) - (GC_{pi})] \text{ (lb/ft}^2\text{)} \tag{Eq. 6-22}$$

The terms in Eq. 6-22 are defined above.

Design wind pressures on C&C for buildings with $h > 60$ ft are determined from the following equation:

$$p = q(GC_p) - q_i(GC_{pi}) \text{ (lb/ft}^2\text{)} \tag{Eq. 6-23}$$

The terms in Eq. 6-23 are defined above.

An alternate procedure (see Section 6.5.12.4.3 of the Standard) to calculate design pressures on C&C for buildings with mean roof height of $60 < h < 90$ ft is to use Eq. 6-22 and associated pressure coefficients. Since this equation, which is for buildings with h equal or less than 60 ft, uses q_h for positive and negative external pressures, the resulting pressures may be higher in some cases.

If a component or cladding element has tributary area (not effective area) greater than 700 ft² (see Section 6.5.12.1.3 of the Standard), it is permitted to be designed using the provisions of MWFRS, Eq. 6-17, and associated pressure coefficients.

2.2.3.2 Design Pressures for Open Buildings

Design wind pressures for the MWFRS and C&C elements of open buildings are determined using the following equation:

$$p = q_h G C_N \qquad \text{(Eqs. 6-25 \& 6-26)}$$

The terms in Eqs. 6-25 and 6-26 are defined above. The net pressure coefficient, C_N, includes the combined pressure on top and bottom surfaces.

2.2.3.3 Design Wind Loads on Other Structures

The design wind-force on other structures is determined by the following equation:

$$F = q_z G C_f A_f \text{ (lb)} \qquad \text{(Eq. 6-28)}$$

The terms of Eq. 6-28 are defined above. The area, A_f, is the exposed area projected on a plane normal to the wind direction unless it is specified with the value of force coefficient, C_f. The force, F, is in the direction of wind except when it is specified with the value of C_f.

2.3 Method 3, Wind Tunnel Procedure

For those situations where the analytical procedure is considered uncertain or inadequate, or where more accurate wind pressures are desired, consideration should be given to wind tunnel tests. The Standard lists a set of conditions in Section 6.6 that must be satisfied for the proper conduct of such tests. The wind tunnel is particularly useful for obtaining detailed information about pressure distributions on complex shapes and the dynamic response of structures. Model scales for structural applications can range from 1:50 for a single-family dwelling to 1:400 for tall buildings. Even smaller scales may be used to model long-span bridges. Of equal importance is the ability to model complex topography at scales of the order of 1:10,000 and assess the effects of features such as hills, mountains, or river gorges on the near-surface winds. Details on wind tunnel modeling for structural or civil engineering applications may be found in Cermak (1977), Reinhold (1982), and ASCE (2006).

2.4 Equations for Graphs

Figures 6-11A through 6-17 of the Standard give external pressure coefficient (GC_p) values for C&C for buildings as a function of effective area of component and cladding. Wind tunnel results found this relationship between pressure coefficients and effective area to be a logarithmic function. The scale of effective area in the figures is a log scale, which makes it very difficult to interpolate. Equations for each of the lines in these figures are given in **Tables G2–1** through **G2–10**. The equations can be used to determine wind loads.

Table G2–1 Walls for Buildings with $h \leq 60$ ft (Figure 6-11A in ASCE 7-05)

Positive: Zones 4 and 5

$(GC_p) = 1.0$	for $A = 10$ ft^2
$(GC_p) = 1.1766 - 0.1766 \log A$	for $10 < A \leq 500$ ft^2
$(GC_p) = 0.7$	for $A > 500$ ft^2

Negative: Zone 4

$(GC_p) = -1.1$	for $A = 10$ ft^2
$(GC_p) = -1.2766 + 0.1766 \log A$	for $10 < A \leq 500$ ft^2
$(GC_p) = -0.8$	for $A > 500$ ft^2

Negative: Zone 5

$(GC_p) = -1.4$	for $A = 10$ ft^2
$(GC_p) = -1.7532 + 0.3532 \log A$	for $10 < A \leq 500$ ft^2
$(GC_p) = -0.8$	for $A > 500$ ft^2

Table G2–2 Gable Roofs with $h \leq 60$ ft, $\theta = 7°$ (Figure 6-11B in ASCE 7-05)

Positive with and without overhang: Zones 1, 2, and 3

$(GC_p) = 0.3$	for $A = 10$ ft²
$(GC_p) = 0.4000 - 0.1000 \log A$	for $10 < A \leq 100$ ft²
$(GC_p) = 0.2$	for $A > 100$ ft²

Negative without overhang: Zone 1

$(GC_p) = -1.0$	for $A = 10$ ft²
$(GC_p) = -1.1000 + 0.1000 \log A$	for $10 < A \leq 100$ ft²
$(GC_p) = -0.9$	for $A > 100$ ft²

Negative without overhang: Zone 2

$(GC_p) = -1.8$	for $A = 10$ ft²
$(GC_p) = -2.5000 + 0.7000 \log A$	for $10 < A \leq 100$ ft²
$(GC_p) = -1.1$	for $A > 100$ ft²

Negative without overhang: Zone 3

$(GC_p) = -2.8$	for $A = 10$ ft²
$(GC_p) = -4.5000 + 1.7000 \log A$	for $10 < A \leq 100$ ft²
$(GC_p) = -1.1$	for $A > 100$ ft²

Negative with overhang: Zones 1 and 2

$(GC_p) = -1.7$	for $A = 10$ ft²
$(GC_p) = -1.8000 + 0.1000 \log A$	for $10 < A \leq 100$ ft²
$(GC_p) = -3.0307 + 0.7153 \log A$	for $100 < A \leq 500$ ft²
$(GC_p) = -1.1$	for $A > 500$ ft²

Negative with overhang: Zone 3

$(GC_p) = -2.8$	for $A = 10$ ft²
$(GC_p) = -4.8000 + 2.0000 \log A$	for $10 < A \leq 100$ ft²
$(GC_p) = -0.8$	for $A > 100$ ft²

Table G2–3 Gable and Hip Roofs with $h \leq 60$ ft, $7° < \theta = 27°$ (Figure 6-11C in ASCE 7-05)

Positive with and without overhang: Zones 1, 2, and 3

$(GC_p) = 0.5$	for $A = 10$ ft^2
$(GC_p) = 0.7000 - 0.2000 \log A$	for $10 < A \leq 100$ ft^2
$(GC_p) = 0.3$	for $A > 100$ ft^2

Negative with and without overhang: Zone 1

$(GC_p) = -0.9$	for $A = 10$ ft^2
$(GC_p) = -1.0000 + 0.1000 \log A$	for $10 < A \leq 100$ ft^2
$(GC_p) = -0.8$	for $A > 100$ ft^2

Negative without overhang: Zones 2

$(GC_p) = -1.7$	for $A = 10$ ft^2
$(GC_p) = -2.2000 + 0.5000 \log A$	for $10 < A \leq 100$ ft^2
$(GC_p) = -1.2$	for $A > 100$ ft^2

Negative without overhang: Zones 3

$(GC_p) = -2.6$	for $A = 10$ ft^2
$(GC_p) = -3.2000 + 0.6000 \log A$	for $10 < A \leq 100$ ft^2
$(GC_p) = -2.0$	for $A > 100$ ft^2

Negative with overhang: Zone 2

$(GC_p) = -2.2$	for all A ft^2

Negative with overhang: Zone 3

$(GC_p) = -3.7$	for $A = 10$ ft^2
$(GC_p) = -4.9000 + 1.2000 \log A$	for $10 < A \leq 100$ ft^2
$(GC_p) = -2.5$	for $A > 100$ ft^2

Table G2–4 Gable Roofs with $h \leq 60$ ft, $27° < \theta \leq 45°$ (Figure 6-11D in ASCE 7-05)

Positive with and without overhang: Zones 1, 2, and 3

$(GC_p) = 0.9$	for $A = 10$ ft^2
$(GC_p) = 1.0000 - 0.1000 \log A$	for $10 < A \leq 100$ ft^2
$(GC_p) = 0.8$	for $A > 100$ ft^2

Negative with and without overhang: Zone 1

$(GC_p) = -1.0$	for $A = 10$ ft^2
$(GC_p) = -1.2000 + 0.2000 \log A$	for $10 < A \leq 100$ ft^2
$(GC_p) = -0.8$	for $A > 100$ ft^2

Negative without overhang: Zones 2 and 3

$(GC_p) = -1.2$	for $A = 10$ ft^2
$(GC_p) = -1.4000 + 0.2000 \log A$	for $10 < A \leq 100$ ft^2
$(GC_p) = -1.0$	for $A > 100$ ft^2

Negative with overhang: Zones 2 and 3

$(GC_p) = -2.0$	for $A = 10$ ft^2
$(GC_p) = -2.2000 + 0.2000 \log A$	for $10 < A \leq 100$ ft^2
$(GC_p) = -1.8$	for $A > 100$ ft^2

Table G2–5   Multispan Gabled Roofs with $h \leq 60$ ft, $10° < \theta \leq 30°$ (Figure 6-13 in ASCE 7-05)

Positive: Zones 1, 2, and 3

$(GC_p) = 0.6$ for $A = 10$ ft^2

$(GC_p) = 0.8000 - 0.2000 \log A$ for $10 < A \leq 100$ ft^2

$(GC_p) = 0.4$ for $A > 100$ ft^2

Negative: Zone 1

$(GC_p) = -1.6$ for $A = 10$ ft^2

$(GC_p) = -1.8000 + 0.2000 \log A$ for $10 < A \leq 100$ ft^2

$(GC_p) = -1.4$ for $A > 100$ ft^2

Negative: Zone 2

$(GC_p) = -2.2$ for $A = 10$ ft^2

$(GC_p) = -2.7000 + 0.5000 \log A$ for $10 < A \leq 100$ ft^2

$(GC_p) = -1.7$ for $A > 100$ ft^2

Negative: Zone 3

$(GC_p) = -2.7$ for $A = 10$ ft^2

$(GC_p) = -3.7000 + 1.0000 \log A$ for $10 < A \leq 100$ ft^2

$(GC_p) = -1.7$ for $A > 100$ ft^2

Table G2–6 Multispan Gable Roofs with $h \leq 60$ ft, $30° < \theta \leq 45°$ (Figure 6-13 in ASCE 7-05)

Positive: Zones 1, 2, and 3

$(GC_p) = 1.0$ for $A = 10$ ft^2

$(GC_p) = 1.2000 - 0.2000 \log A$ for $10 < A \leq 100$ ft^2

$(GC_p) = 0.8$ for $A > 100$ ft^2

Negative: Zone 1

$(GC_p) = -2.0$ for $A = 10$ ft^2

$(GC_p) = -2.9000 + 0.9000 \log A$ for $10 < A \leq 100$ ft^2

$(GC_p) = -1.1$ for $A > 100$ ft^2

Negative: Zone 2

$(GC_p) = -2.5$ for $A = 10$ ft^2

$(GC_p) = -3.3000 + 0.8000 \log A$ for $10 < A \leq 100$ ft^2

$(GC_p) = -1.7$ for $A > 100$ ft^2

Negative: Zone 3

$(GC_p) = -2.6$ for $A = 10$ ft^2

$(GC_p) = -3.5000 + 0.9000 \log A$ for $10 < A \leq 100$ ft^2

$(GC_p) = -1.7$ for $A > 100$ ft^2

Table G2–7   Monoslope Roofs with $h \leq 60$ ft, $3° < \theta \leq 10°$ (Figure 6-14A in ASCE 7-05)

Positive: All Zones

$(GC_p) = 0.3$	for $A = 10$ ft^2
$(GC_p) = 0.4000 - 0.1000 \log A$	for $10 < A \leq 100$ ft^2
$(GC_p) = 0.2$	for $A > 100$ ft^2

Negative: Zone 1

$(GC_p) = -1.1$	for all A ft^2

Negative: Zone 2

$(GC_p) = -1.3$	for $A = 10$ ft^2
$(GC_p) = -1.4000 + 0.1000 \log A$	for $10 < A \leq 100$ ft^2
$(GC_p) = -1.2$	for $A > 100$ ft^2

Negative: Zone 2'

$(GC_p) = -1.6$	for $A = 10$ ft^2
$(GC_p) = -1.7000 + 0.1000 \log A$	for $10 < A \leq 100$ ft^2
$(GC_p) = -1.5$	for $A > 100$ ft^2

Negative: Zone 3

$(GC_p) = -1.8$	for $A = 10$ ft^2
$(GC_p) = -2.4000 + 0.6000 \log A$	for $10 < A \leq 100$ ft^2
$(GC_p) = -1.2$	for $A > 100$ ft^2

Negative: Zone 3'

$(GC_p) = -2.6$	for $A = 10$ ft^2
$(GC_p) = -3.6000 + 1.0000 \log A$	for $10 < A \leq 100$ ft^2
$(GC_p) = -1.6$	for $A > 100$ ft^2

Table G2–8   Monoslope Roofs with $h \leq 60$ ft, $10° < \theta \leq 30°$ (Figure 6-14B in ASCE 7-05)

Positive: All Zones

$(GC_p) = 0.4$	for $A = 10$ ft^2
$(GC_p) = 0.5000 - 0.1000 \log A$	for $10 < A \leq 100$ ft^2
$(GC_p) = 0.3$	for $A > 100$ ft^2

Negative: Zone 1

$(GC_p) = -1.3$	for $A = 10$ ft^2
$(GC_p) = -1.5000 + 0.2000 \log A$	for $10 < A \leq 100$ ft^2
$(GC_p) = -1.1$	for $A > 100$ ft^2

Negative: Zone 2

$(GC_p) = -1.6$	for $A = 10$ ft^2
$(GC_p) = -2.0000 + 0.4000 \log A$	for $10 < A \leq 100$ ft^2
$(GC_p) = -1.2$	for $A > 100$ ft^2

Negative: Zone 3

$(GC_p) = -2.9$	for $A = 10$ ft^2
$(GC_p) = -3.8000 + 0.9000 \log A$	for $10 < A \leq 100$ ft^2
$(GC_p) = -2.0$	for $A > 100$ ft^2

Table G2–9 Sawtooth Roofs with $h \leq 60$ ft, $\theta \leq 10°$ (Figure 6-15 in ASCE 7-05)

Positive: Zone 1

$(GC_p) = 0.7$ for $A = 10$ ft^2

$(GC_p) = 0.8766 - 0.1766 \log A$ for $10 < A \leq 500$ ft^2

$(GC_p) = 0.4$ for $A > 500$ ft^2

Positive: Zone 2

$(GC_p) = 1.1$ for $A = 10$ ft^2

$(GC_p) = 1.4000 - 0.3000 \log A$ for $10 < A \leq 100$ ft^2

$(GC_p) = 0.8$ for $A > 100$ ft^2

Positive: Zone 3

$(GC_p) = 0.8$ for $A = 10$ ft^2

$(GC_p) = 0.9000 - 0.1000 \log A$ for $10 < A \leq 100$ ft^2

$(GC_p) = 0.7$ for $A > 100$ ft^2

Negative: Zone 1

$(GC_p) = -2.2$ for $A = 10$ ft^2

$(GC_p) = -2.8474 + 0.6474 \log A$ for $10 < A \leq 500$ ft^2

$(GC_p) = -1.1$ for $A > 500$ ft^2

Negative: Zone 2

$(GC_p) = -3.2$ for $A = 10$ ft^2

$(GC_p) = -4.1418 + 0.9418 \log A$ for $10 < A \leq 500$ ft^2

$(GC_p) = -1.6$ for $A > 500$ ft^2

Negative: Zone 3 (span A)

$(GC_p) = -4.1$ for $A = 10$ ft^2

$(GC_p) = -4.5000 + 0.4000 \log A$ for $10 < A \leq 100$ ft^2

$(GC_p) = -8.2782 + 2.2891 \log A$ for $100 < A \leq 500$ ft^2

$(GC_p) = -2.1$ for $A > 500$ ft^2

Negative: Zone 3 (spans B, C, D)

$(GC_p) = -2.6$ for $A = 100$ ft^2

$(GC_p) = -4.6030 + 1.0015 \log A$ for $100 < A \leq 500$ ft^2

$(GC_p) = -1.9$ for $A > 500$ ft^2

Table G2–10 Roof and Walls for Buildings with $h > 60$ ft (Figure 6-17 in ASCE 7-05)

Roofs $\theta = 10°$

Negative: Zone 1

$(GC_p) = -1.4$	for $A = 10$ ft^2
$(GC_p) = -1.6943 + 0.2943 \log A$	for $10 < A \leq 500$ ft^2
$(GC_p) = -0.9$	for $A > 500$ ft^2

Negative: Zone 2

$(GC_p) = -2.3$	for $A = 10$ ft^2
$(GC_p) = -2.7120 + 0.4120 \log A$	for $10 < A \leq 500$ ft^2
$(GC_p) = -1.6$	for $A > 500$ ft^2

Negative: Zone 3

$(GC_p) = -3.2$	for $A = 10$ ft^2
$(GC_p) = -3.7297 + 0.5297 \log A$	for $10 < A \leq 500$ ft^2
$(GC_p) = -2.3$	for $A > 500$ ft^2

Walls All θ

Positive: Zones 4 and 5

$(GC_p) = 0.9$	for $A = 20$ ft^2
$(GC_p) = 1.1792 - 0.2146 \log A$	for $20 < A \leq 500$ ft^2
$(GC_p) = 0.6$	for $A > 500$ ft^2

Negative: Zone 4

$(GC_p) = -0.9$	for $A = 20$ ft^2
$(GC_p) = -1.0861 + 0.1431 \log A$	for $20 < A \leq 500$ ft^2
$(GC_p) = -0.7$	for $A > 500$ ft^2

Negative: Zone 5

$(GC_p) = -1.8$	for $A = 20$ ft^2
$(GC_p) = -2.5445 + 0.5723 \log A$	for $20 < A \leq 500$ ft^2
$(GC_p) = -1.0$	for $A > 500$ ft^2

Chapter 3

Examples

In this chapter, 13 examples illustrate how wind loads are determined using the simplified and analytical procedures described in ASCE 7-05. These examples provide guidance to the user of the Standard in determining wind loads for several types of buildings.

These examples represent a variety of situations in determination of wind loads. The equation, table, figure, and section numbers of ASCE 7-05 are cited where appropriate. Every effort has been made to check the accuracy of the numbers in calculations, although no absolute assurance is given.

3.1 Example 1: 30-ft × 60-ft × 15-ft Commercial Building with Concrete Masonry Unit Walls

In this example, design wind pressures for a typical load-bearing one-story masonry building are determined. The building is shown in **Figure G3–1**, and data are as shown here.

This example uses Method 2, Analytical Procedure, of Section 6.5 of ASCE 7-05 for rigid buildings of all heights. The same building is illustrated in Ex. 2 (Section 3.2) using Method 1, Simplified Procedure, of Section 6.4 of ASCE 7-05.

Example No.	Building/Methodology	Section No.	Figure No. (This Guide)
1	30-ft × 60-ft × 15-ft commercial building with concrete masonry unit (CMU) walls	3.1	G3-1
2	Commercial building from Ex. 1 using simplified procedure	3.2	G3-1
3	100-ft × 200-ft × 160-ft-high office building located in hurricane zone	3.3	G3-6
4	Office building from Ex. 3 located on an escarpment	3.4	G3-10
5	A typical 2,500-ft^2 house with gable/hip roof	3.5	G3-11(a)–(d)
6	200-ft × 250-ft gable roof commercial/warehouse building using all height provisions	3.6	G3-13
7	Commercial/warehouse building from Ex. 6 using low-rise building provisions	3.7	G3-13
8	40-ft × 80-ft commercial building with monoslope roof with overhang	3.8	G3-27
9	U-shaped apartment building	3.9	G3-33
10	50-ft × 20-ft billboard sign on poles (flexible) 60 ft above ground	3.10	G3-38
11	Domed-roof building	3.11	G3-40
12	Unusually shaped building	3.12	G3-43
13	Open roof	3.13	G3-50

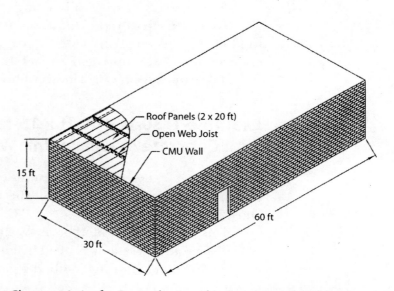

Figure G3–1   Building Characteristics for Examples 1 and 2, Commercial Building

Location	Corpus Christi, Texas
Topography	Homogeneous
Terrain	Flat, open terrain
Dimensions	30 ft × 60 ft × 15 ft, flat roof
Building Use	Shop
Framing	CMU walls on three sides Steel framing in front with glass Open web joists, 30-ft span spaced at 5 ft on center, covered with metal panel to provide roof diaphragm action
Cladding	Roof metal panels are 2-ft wide, 20-ft long Doors and glass size vary; glass is debris resistant
Roof top equipment	1 – 10 ft (face normal to wind) × 5 ft × 5 ft high air handler 1 – 3 ft diameter smooth surface vent pipe

3.1.1 Basic Wind Speed

Selection of the basic wind speed is addressed in Section 6.5.4 of the Standard. Basic wind speed for Corpus Christi, Texas, is 130 mph (Figure 6-1a of the Standard).

3.1.2 Exposure

The building is located on flat and open terrain. It does not fit Exposures B or D, therefore use Exposure C (Sections 6.5.6.2 and 6.5.6.3 of the Standard).

3.1.3 Building Classification

The building function is shops. It is not considered an essential facility. Building Category II is appropriate; see Table 1-1 of the Standard.

3.1.4 Velocity Pressure

The velocity pressures are computed using

$$q_z = 0.00256\, K_z K_{zt} K_d V^2 I \qquad \text{(Eq. 6-15)}$$

where

K_z = 0.85 from Table 6-3 of the Standard for Case 1 (C&C) and Case 2 (MWFRS); for 0 to 15 ft, there is only one value: $K_z = K_h$

K_{zt} = 1.0 for homogeneous topography (see Section 6.5.7 of the Standard)

K_d = 0.85 for buildings (see Table 6-4 of the Standard)

V = 130 mph (see Figure 6-1a of the Standard)

I = 1.0 for Category II building (see Table 6-1 of the Standard) therefore,

$$q_z = 0.00256 \, (0.85) \, (1.0) \, (0.85) \, (130)^2 \, (1.0) = 31.3 \text{ psf}$$

$$q_h = 31.3 \text{ psf for } h = 15 \text{ ft}$$

Gust Effect Factor

The building is considered a rigid structure. Section 6.5.8.1 of the Standard permits use of $G = 0.85$. If the detailed procedure for a rigid structure is used (Section 6.5.8.1 of the Standard), the calculated value of $G = 0.89$; however, the Standard permits the use of the value of $G = 0.85$. Detailed calculations for G value are illustrated in Ex. 3 (Section 3.3 of this guide).

Use $G = 0.85$ for this example.

Internal Pressure Coefficient

The building is located in a hurricane-prone area (see definition of wind-borne debris region in Section 6.2 of the Standard). Section 6.5.9.3 requires that glazing be considered openings unless protected or debris resistant. The example building has debris-resistant glazing, and other openings are such that it does not qualify as a partially enclosed or open building.

Use $(GC_{pi}) = +0.18$ and -0.18 for enclosed buildings (see Figure 6-5 of the Standard).

3.1.5 Design Wind Pressures for MWFRS

Design wind pressures are determined using

$$p = q \, G \, C_p - q_i \, (GC_{pi}) \qquad \text{(Eq. 6-17)}$$

where

$q = q_z$ for windward wall (31.3 psf for this example)

$q = q_h$ for leeward wall, side walls, and roof (31.3 psf for this example)

$G = 0.85$

C_p = Values of external pressure coefficients

$q_i = q_h$ for enclosed building (31.3 psf)

$(GC_{pi}) = +0.18$ and -0.18

The values of external pressure coefficients are obtained from Figure 6-6 of the Standard.

Wall Pressure Coefficient

The windward wall pressure coefficient is 0.8.

The side wall pressure coefficient is -0.7.

The leeward wall pressure coefficients are a function of L/B ratio.

For $L/B = 0.5$, $C_p = -0.5$ for wind normal to 60 ft

For $L/B = 2.0$, $C_p = -0.3$ for wind normal to 30 ft

Roof Pressure Coefficient

The roof pressure coefficients are a function of roof slope and h/L. For $\theta < 10°$ and $h/L = 0.25$ and 0.5,

First value

$C_p = -0.9$ for distance 0 to h

$C_p = -0.5$ for distance h to $2h$

$C_p = -0.3$ for distance $>2h$

Second value

$C_p = -0.18$ for distance 0 to end.

This value of smaller uplift pressures on the roof can become critical when wind load is combined with roof live load or snow load; load combinations are given in Section 2.3 and 2.4 of the Standard. For brevity, loading for this value is not shown in this example.

MWFRS Pressures

Windward wall

$p = 31.3 \, (0.85) \, (0.8) - 31.3 \, (\pm 0.18) = 21.3 \pm 5.6$ psf

Leeward wall

$p = 31.3 \, (0.85) \, (-0.5) - 31.3 \, (\pm 0.18) = -13.3 \pm 5.6$ psf for wind normal to 60 ft

$p = 31.3 \, (0.85) \, (-0.3) - 31.3 \, (\pm 0.18) = -8.0 \pm 5.6$ psf for wind normal to 30 ft

Roof – First value

$p = 31.3 \, (0.85) \, (-0.9) - 31.3 \, (\pm 0.18)$
  $= -23.9 \pm 5.6$ psf   for 0 to 15 ft
  $= -13.3 \pm 5.6$ psf   for 15 to 30 ft
  $= -8.0 \pm 5.6$ psf   for > 30 ft

The MWFRS design pressures for two directions are shown in **Figures G3–2** and **G3–3**. The internal pressures shown are to be added to the external pressures as appropriate. The internal pressures of the same sign act on all surfaces; thus, they cancel out for total horizontal shear.

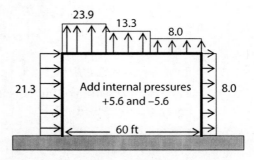

Figure G3–2   Design Pressures for MWFRS When Wind Is Normal to 30-ft Wall

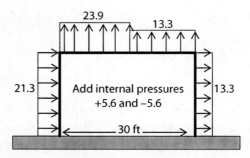

Figure G3–3   Design Pressures for MWFRS When Wind Is Normal to 60-ft Wall

3.1.6 Design Wind Load Cases

According to Section 6.5.12.3 of the Standard, this building shall be designed for the wind load Cases 1 and 3 as defined in Figure 6-9. Load Case 1 has been considered above. **Figure G3–4** is for Load Case 3 where the windward and leeward pressures are taken as 75% of the specified values.

3.1.7 Design Pressures for C&C

Design wind pressures are determined using the equation

$$p = q_h\,[\,(GC_p) - (GC_{pi})\,] \tag{Eq. 6-22}$$

where

q_h = 31.3 psf

(GC_p) = Values obtained from Figure 6-11 of the Standard; they are a function of effective area and zone

(GC_{pi}) = +0.18 and −0.18

Wall Pressures

CMU walls are supported at the roof diaphragm and at ground, span = 15 ft. CMU wall effective wind area is determined using the definition from Section 6.2 of the Standard: "the width of effective area need not be less than one-third of the span."

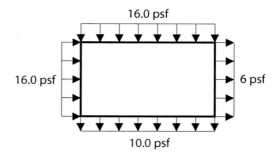

Figure G3–4 Load Case 3

CMU wall effective wind area, $A = 15 \, (15/3) = 75 \text{ ft}^2$

In Figure 6-11A of the Standard, Note 5 suggests that the pressure coefficient values for walls can be reduced by 10% for roof slope of 10° or less thus calculations below take 90% of the graphed values. The values of (GC_p) are obtained from the figure or from equations of the graphs (see Section 2.4 of this guide).

Corner Zone 5 distance

smaller of $a = 0.1 \, (30) = 3$ ft (controls)
or $a = 0.4 \, (15) = 6$ ft

Corner Zone 5

$p = 31.3 \, [(-1.09) \, (0.9) - (\pm 0.18)] = -36.3$ psf

$p = 31.3 \, [(0.85) \, (0.9) - (\pm 0.18)] = +29.6$ psf

Interior Zone 4

$p = 31.3 \, [(-0.95) \, (0.9) - (\pm 0.18)] = -32.4$ psf

$p = 31.3 \, [(0.85) \, (0.9) - (\pm 0.18)] = +29.6$ psf

Note: The CMU walls have uplift pressure from the roof, which is determined on the basis of MWFRS.

Pressure for glazing and mullions can be determined similarly with the known effective wind area.

Roof Joist Pressures

Roof joists span 30 ft and are spaced 5 ft apart. The joist can be in Zone 1 (interior of roof) or Zone 2 (eave area). Zone 3 (roof corner area) acts only on a part of the joist.

Width of Zones 2 and 3 (Figure 6-11B)

smaller of $a = 0.1 \, (30) = 3$ ft (controls)
or $a = 0.4 \, (15) = 6$ ft

Joist Effective Wind Area

larger of $\quad A = 30 \times 5 = 150$ ft

or $\quad A = 30 \times (30/3) = 300$ ft (controls)

The values of (GC_p) are obtained from Figure 6-11B of the Standard or from equations of the graphs using effective area $A = 300$ ft^2.

Interior Zone 1

$p = 31.3 \,[-0.9 \pm 0.18] = -33.8$ psf

$p = 31.3 \,[+0.2 \pm 0.18] = +11.9$ psf

Eave Zone 2 and Corner Zone 3

$p = 31.3 \,[-1.1 \pm 0.18] = -40.1$ psf

$p = 31.3 \,[+0.2 \pm 0.18] = +11.9$ psf

Roof Panel Pressures

Even though roof panel length is 20 ft, each panel spans 5 ft between joists.

Roof Panel Effective Area

larger of $\quad A = 5 \times 2 = 10$ ft^2 (controls)

or $\quad A = 5 \times (5/3) = 8$ ft^2 (width of Zones 2 and 3, $a = 3$ ft)

Interior Zone 1

$p = 31.3 \,[-1.0 \pm 0.18] = -36.9$ psf

$p = 31.3 \,[+0.3 \pm 0.18] = +15.0$ psf

Eave Zone 2

$p = 31.3 \,[-1.8 \pm 0.18] = -62.0$ psf

$p = 31.3 \,[+0.3 \pm 0.18] = +15.0$ psf

Corner Zone 3

$p = 31.3 \,[-2.8 \pm 0.18] = -93.3$ psf

$p = 31.3 \,[+0.3 \pm 0.18] = +15.0$ psf

Notes:

- Internal pressure coefficient of +0.18 or −0.18 is used to give critical pressures.
- The roof panel fasteners design pressures will be the same as metal panel since values of (GC_p) are the same for wind effective areas less than 10 ft^2.

3.1.8 Design Forces on Roof Top Equipment

Design wind forces are determined using the equation

$$F = 1.9 q_z G C_f A_f \qquad \text{(Eq. 6-28)}$$

where

- q_z = 31.3 psf
- G = gust effect factor from Section 6.5.8
- C_f = force coefficients from Figure 6-21
- A_f = projected area normal to the wind
- 1.9 = magnification factor when $A_f < 0.1Bh$ from Section 6.5.15.1. Note if $A_f \geq Bh$, the magnification factor is 1.0. It may be linearly interpolated from 1.9 to 1.0 as the value of A_f increases from $0.1Bh$ to Bh.

Rectangular air handler

From Figure 6-21, for rectangular (square parameters used since rectangular is not a noted shape), h/D = 15 ft (height of structure)/5 ft (least dimension of square cross-section) = 3.

Interpolating from Figure 6-21, $h/D = 3$, $C_f = 1.33$.

A_f = 10 ft × 5 ft = 50 sq ft

A_f must be compared to $0.1Bh$ in order to determine the magnitude of the magnification factor. B = the face of the building normal to the wind, therefore if the windward wall is 60 ft, $0.1Bh = 0.1 (60)(15) = 90$ sq ft, therefore $A_f < 0.1Bh$ and the magnification factor = 1.9.

Thus, F = 31.3 psf (1.9) (0.85) (1.33) (50 sq ft) = 3,361 lbs.

Round vent pipe

From Figure 6-21, for a round cross-section, $D/\sqrt{q_z}$ where D = diameter of round cross-section and q_z = velocity pressure at height z, or $3/\sqrt{31.3} = 0.54$, which is less than 2.5

With $h/D = 3$, $C_f = 0.73$ interpolating in Figure 6-21.

A_f = 3 ft × 10 ft = 30 sq ft.

$A_f < 0.1Bh$, therefore the magnification factor = 1.9.
Thus, F = 31.3 psf (1.9) (0.85) (0.73) (30 sq ft) = 1,108 lb.

3.2 Example 2: Ex. 1 Using Simplified Procedure

In this example, design wind pressures for the building of Ex. 1 are determined using the simplified procedure of Section 6.4 of the Standard. Data for the building are the same as Ex. 1 (see Section 3.1 and **Figure G3–1**).

In order to use the simplified procedure, all conditions of Section 6.4.1 of the Standard must be satisfied, namely:

1. It is a simple diaphragm building.
2. Its mean roof height h is less than 60 ft and does not exceed the least horizontal dimension.
3. Since the building has debris-resistant glazing and no dominant opening in any one wall, it can be classified as an enclosed building. It also conforms to the wind-borne debris provisions of Section 6.5.9.3 of the Standard.
4. It has a regular shape.
5. It is a rigid building (h/width << 4) (see Commentary in the Standard).
6. There is no expansion joint.
7. There is no abrupt change in topography (see Section 6.5.7.1 of the Standard for requirements of topographic effects).
8. It has an approximately symmetrical cross section in each direction with a flat roof.

Wind pressures for both the MWFRS and C&C can be obtained using the simplified procedure.

3.2.1 Basic Wind Speed

Basic wind speed for Corpus Christi, Texas, is 130 mph (see Figure 6-1a of the Standard).

3.2.2 Building Classification

Building Category II is appropriate. Importance Factor $I = 1.00$ (see Table 6-1 of the Standard).

3.2.3 Exposure

The building is located on flat and open terrain. It does not fit Exposures B or D; therefore, use Exposure C (Sections 6.5.6.2 and 6.5.6.3 of the Standard). Note that wind pressure values given in Figures 6-2 and 6-3 of the Standard are for Exposure B.

3.2.4 Height and Exposure Adjustment Coefficient λ

From Figure 6-2 of the Standard, $\lambda = 1.21$.

3.2.5 Design Wind Pressures for MWFRS

This building has a flat roof, so only Load Case 1 is checked (see **Table G3–1**)

$$p = \lambda\, k_{zt}\, I\, p_{s30} = 1.21 \times 1.0 \times 1.0 \times p_{s30} \qquad \text{(Eq. 6-1)}$$

Table G3-1 Design Wind Pressures

Zones	A	C	E	F	G	H
p_s (psf)	32.4	21.5	−39.0	−22.1	−27.1	−17.2

Notes: (1) Zones are defined in Figure 6-2 of the Standard: a = smaller of $0.1 \times 30 = 3$ ft (control) or $0.4 \times 15 = 6$ ft. (2) The load patterns shown in **Figure G3–5** shall be applied to each corner of the building in turn as the reference corner.

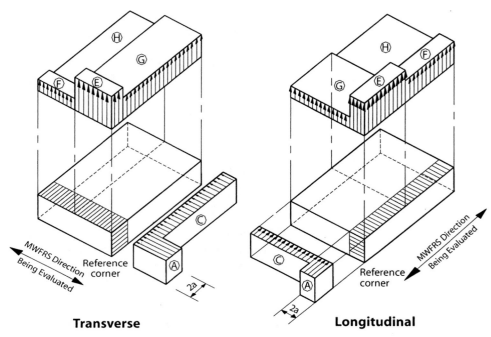

Figure G3–5 *Design Wind Pressure for Transverse and Longitudinal Directions*

In the simplified procedure, design roof pressure includes internal pressure. The wall pressure is the combined windward and leeward wall pressures (internal pressure cancels).

3.2.6 Design Pressures for C&C

According to Section 6.4.2.2 of the Standard,

$$p_{net} = \lambda\, k_{zt}\, I\, p_{net30} = 1.21 \times 1.0 \times 1.0 \times p_{net30} \qquad \text{(Eq. 6-2)}$$

Wall Pressures

The effective wind area for a CMU wall is 75 ft² (see Ex. 1). Linear interpolation is permitted in Figure 6-3 of the Standard.

Zone 4

$$p_{net} = 1.21 \times 1.0 \times 1.0 \times 26.6 = 32.2 \text{ psf}$$
$$p_{net} = 1.21 \times 1.0 \times 1.0 \times (-29.1) = -35.2 \text{ psf}$$

Zone 5

$$p_{net} = 1.21 \times 1.0 \times 1.0 \times 26.6 = 32.2 \text{ psf}$$
$$p_{net} = 1.21 \times 1.0 \times 1.0 \times (-33.0) = -39.9 \text{ psf}$$

Roof Joist Pressures

From Figure 6-3 of the Standard, for $V = 130$ mph, for effective wind area of 300 ft^2, the design pressures are calculated as follows.

Zone 1

$$p_{net} = 1.21 \times 1.0 \times 1.0 \times 9.8 = 11.9 \text{ psf}$$
$$p_{net} = 1.21 \times 1.0 \times 1.0 \times (-27.8) = -33.6 \text{ psf}$$

Zones 2 and 3

$$p_{net} = 1.21 \times 1.0 \times 1.0 \times 9.8 = 11.9 \text{ psf}$$
$$p_{net} = 1.21 \times 1.0 \times 1.0 \times (-33.0) = -39.9 \text{ psf}$$

Roof Panel Pressures

Effective wind area for roof panel is 10 ft^2 (see Ex. 1). From Figure 6-3 of the Standard, for $V = 130$ mph, for effective wind area of 10 ft^2, the design pressures are calculated as follows.

Zone 1

$$p_{net} = 1.21 \times 1.0 \times 1.0 \times 12.4 = 15.0 \text{ psf}$$
$$p_{net} = 1.21 \times 1.0 \times 1.0 \times (-30.4) = -36.8 \text{ psf}$$

Zone 2

$$p_{net} = 1.21 \times 1.0 \times 12.4 = 15.0 \text{ psf}$$
$$p_{net} = 1.21 \times 1.0 \times (-51.0) = -61.7 \text{ psf}$$

Zone 3

$$p_{net} = 1.21 \times 1.0 \times 12.4 = 15.0 \text{ psf}$$
$$p_{net} = 1.21 \times 1.0 \times (-76.8) = -92.9 \text{ psf}$$

The analytical procedure in Ex. 1 yields C&C design pressures close to the results of the simplified procedure.

3.3 Example 3: 100-ft × 200-ft × 160-ft High Office Building

This building is illustrated in **Figure G3–6**; data for the building is as follows:

Location Near Houston, Texas

Topography Homogeneous

Terrain Suburban

Dimensions 100 ft × 200 ft in plan
Roof height of 157 ft with 3-ft parapet
Flat roof

Framing Reinforced concrete rigid frame in both directions
Floor and roof slabs provide diaphragm action
Fundamental natural frequency is greater than 1 Hz
(Since the height to least horizontal dimension is less than 4, the fundamental frequency is judged to be greater than 1 Hz.)

Cladding Mullions for glazing panels span 11 ft between floor slabs
Mullion spacing is 5 ft

Glazing panels are 5-ft wide × 5-ft 6 in. high (typical); they are wind-borne debris impact resistant in the bottom 60 ft as required by Section 6.5.9.3 of the Standard

The analytical procedure of ASCE 7-05 is to be used.

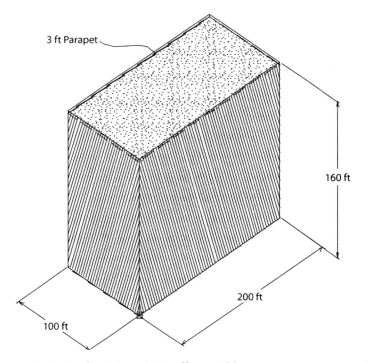

Figure G3–6   Building Characteristics for Example 3, Office Building

3.3.1 Exposure

The building is located in a suburban area; according to Section 6.5.6.3 of the Standard, Exposure B is used.

3.3.2 Building Classification

The building function is office space. It is not considered an essential facility or likely to be occupied by 300 persons in a single area at one time. Therefore, building Category II is appropriate (see Table 1-1 of the Standard).

3.3.3 Basic Wind Speed

Selection of the basic wind speed is addressed in Section 6.5.4 of the Standard. Vicinity of Houston, Texas, is located on the 120-mph contour. The basic wind speed $V = 120$ mph (see Figure 6-1a of the Standard

3.3.4 Velocity Pressures

The velocity pressures are computed using the following equation:

$$q_z = 0.00256\, K_z K_{zt} K_d V^2 I \text{ psf} \qquad \text{(Eq. 6-15)}$$

where

K_z = value obtained from Table 6-3, Case 1 for C&C and Case 2 for MWFRS

K_{zt} = 1.0 for homogeneous topography

K_d = 0.85 for buildings (see Table 6-4 of the Standard)

V = 120 mph

I = 1.0 for Category II classification (see Table 6-1 of the Standard)

$q_z = 0.00256 K_z (1.0)(0.85)(120)^2 (1.0)$

$q_z = 31.3\, K_z$ psf,

Values for K_z and the resulting velocity pressures are given in **Table G3–2**. The velocity pressure at mean roof height, q_h, is 35.1 psf.

3.3.5 Design Wind Pressures for the MWFRS

The design pressures for this building are obtained by equation,

$$p = qGC_p - q_i(GC_{pi}) \qquad \text{(Eq. 6-17)}$$

where

$q = q_z$ for windward wall at height z above ground

$q = q_h$ for leeward wall, side walls, and roof

Table G3–2 q_z Velocity Pressures

	MWFRS		C&C	
Height, ft	K_z	q_z, psf	K_z	q_z, psf
0–15	0.57	17.9	0.70	21.9
30	0.70	21.9	0.70	21.9
50	0.81	25.4	0.81	25.4
80	0.93	29.1	0.93	29.1
120	1.04	32.6	1.04	32.6
Roof = 157	1.12	35.1	1.12	35.1
Parapet = 160	1.13	35.4	1.13	35.4

Note: q_h = 35.1 psf.

q_i = q_h for windward walls, side walls, leeward walls, and roofs for negative internal pressure evaluation in partially enclosed building

q_i = q_z for positive internal pressure evaluation in partially enclosed buildings where height z is defined as the level of the highest opening in the building that could affect the positive internal pressure. For buildings sited in wind-borne debris regions, glazing that is not impact resistant or protected shall be treated as an opening in accordance with Section 6.5.9.3.

G = Gust effect factor for rigid building and structure

C_p = External pressure coefficient

(GC_{pi}) = Internal pressure coefficient

Gust Effect Factor, G

Dimensions of this building where h/least width = 1.6 < 4.0 indicates that it is a rigid structure:

$$G = 0.925 \left\{ \frac{(1+1.7\, g_Q I_{\bar{z}} Q)}{(1+1.7\, g_V I_{\bar{z}})} \right\} \quad \text{(Eq. 6-4)}$$

$g_Q = g_V = 3.4$ \hfill (Sec. 6.5.8.1)

$\bar{z} = 0.6(157) = 94.2$ ft (controls) \hfill (Sec. 6.5.8.1)

$\bar{z} = z_{min} = 30$ ft \hfill (Table 6-2)

$c = 0.30$ \hfill (Table 6-2)

$$I_{\bar{z}} = c\left(\frac{33}{\bar{z}}\right)^{\frac{1}{6}} = 0.30\left(\frac{33}{94.2}\right)^{\frac{1}{6}} = 0.25 \qquad \text{(Eq. 6-5)}$$

$$L_{\bar{z}} = l\left(\frac{\bar{z}}{33}\right)^{\bar{\epsilon}} = 320\left(\frac{94.2}{33}\right)^{\frac{1}{3}} = 454 \text{ ft} \qquad \text{(Eq. 6-7)}$$

$$Q = \sqrt{\frac{1}{1 + 0.63\left(\frac{B+h}{L_{\bar{z}}}\right)^{0.63}}} \qquad \text{(Eq. 6-6)}$$

$B = 100$ ft (smaller value gives larger G)

$$Q = \sqrt{\frac{1}{1 + 0.63\left(\frac{100+157}{454}\right)^{0.63}}} = 0.83$$

$$G = 0.925\left\{\frac{(1 + 1.7 \times 3.4 \times 0.25 \times 0.83)}{(1 + 1.7 \times 3.4 \times 0.25)}\right\} = 0.83$$

Wall External Pressure Coefficients, C_p

The values for the external pressure coefficients for the various wall surfaces are obtained from Figure 6-6 of the Standard and show in **Table G3–3**. The windward wall pressure coefficient is 0.8. The side wall pressure coefficient is –0.7.

The leeward wall pressure coefficient is a function of the L/B ratio. For wind normal to the 200-ft face, $L/B = 100/200 = 0.5$; therefore, the leeward wall pressure coefficient is –0.5. For wind normal to 100-ft face, $L/B = 200/100 = 2.0$; therefore, the leeward wall pressure coefficient is –0.3.

Table G3–3   Wall C_p for Ex. 3

Surface	Wind Direction	L/B	C_p
Windward wall	All	All	0.80
Leeward wall	⊥ to 200-ft face	0.5	–0.50
	∥ to 200-ft face	2.0	–0.30
Side wall	All	All	–0.70

Roof C_p with the wind normal to the 200-ft face

Roof C_p with the wind normal to the 200-ft face are shown in **Table G3–4**. For $h/L = 157/100 \approx 1.6 > 1.0$, and $\theta < 10°$, two zones are specified in Figure 6-6 of the Standard.

First value

    0 to $h/2$ $C_p = -1.3$
    $> h/2$ $C_p = -0.7$

Second value

    $C_p = -0.18$. This value of smaller uplift pressures on the roof can become critical when wind load is combined with roof live load or snow load; load combinations are given in Section 2.3 and 2.4 of the Standard. For brevity, loading for this value is not shown in this example.

    The $C_p = -1.3$ may be reduced with the area over which it is applicable.

    Area = $200 \times 79 = 15{,}800$ ft^2

    Reduction factor = 0.8

    Reduced $C_p = 0.8 \times (-1.3) = -1.04$

Roof C_p with the wind normal to 100-ft face

Roof C_p with the wind normal to the 100-ft face are shown in **Table G3–5**. For $h/L = 157/200 \approx 0.8$, interpolation in Figure 6-6 of the Standard is required.

Roof Calculation for 0 to 79 ft (h/2) from Edge (Wind Normal to 200-ft Face)

    External pressure = $35.1(0.83)(-1.04) = -30.3$ psf

Table G3–4   Roof C_p for Wind Normal to 200-ft Face

Distance from Leading Edge	C_p
0 to $h/2$	−1.04
$> h/2$	−0.70

Note: $h = 157$ ft.

Table G3–5   Roof C_p for Wind Normal to 100-ft Face

Distance from windward edge	$h/L \leq 0.5$	$h/L = 0.8$	$h/L \geq 1.0$
0 to $h/2$	−0.9	−0.98	−1.04
$h/2$ to h	−0.9	−0.78	−0.7
h to $2h$	−0.5	−0.62	−0.7

Roof Calculation for 79 (h/2) to 100-ft from Edge (Wind Normal to 200-ft Face)

External pressure = 35.1 (0.83) (−0.70) = −20.4 psf

External pressures are summarized in **Tables G3–6** and **G3–7**.

Internal Pressure Coefficients, (GC_{pi})

The building is in a hurricane-prone and wind-borne debris region. The glazing is required to be debris resistant up to 60 ft above the ground (Section 6.5.9.3 of the Standard). Above 60 ft, glazing is considered as an opening. In addition, if there is a debris source, such as an aggregate surfaced roof within 1500 ft of the subject building, glazing that is up to 30 ft above the aggregate surfaced roof must be debris impact resistant from the ground to 30 ft above the adjacent building roof. Since there is no information in this example about an adjacent aggregate surfaced roof, the lower 60 ft of the building could be designed as an enclosed building using $GC_{pi} = \pm 0.18$ because of impact resistant glazing. The remainder of the building is classified as a partially enclosed building.

MWFRS Pressures

$$p = qGC_p - q_i(GC_{pi}) \tag{Eq. 6-17}$$

For enclosed buildings

$$GC_{pi} = \pm 0.18$$

Table G3–6 External Pressures for MWFRS: Wind Normal to 200-ft Face

Surface	z (ft)	q (psf)	C_p	External Pressures (psf)
Windward wall	0–15	17.9	0.80	11.9
	30	21.9	0.80	14.5
	50	25.4	0.80	16.9
	80	29.1	0.80	19.3
	120	32.6	0.80	21.7
	157	35.1	0.80	23.3
Leeward wall	All	35.1	−0.50	−14.6
Side walls	All	35.1	−0.70	−20.4
Roof	0–79	35.1	−1.04	−30.3
	79–100	35.1	−0.70	−20.4

Note: $q_h = 35.1$ psf; $G = 0.83$.

Table G3-7 External Pressures for MWFRS: Wind Normal to 100-ft Face

Surface	z (ft)	q (psf)	C_p	External Pressure (psf)
Windward wall	0–15	17.9	0.80	11.9
	30	21.9	0.80	14.5
	50	25.4	0.80	16.9
	80	29.1	0.80	19.3
	120	32.6	0.80	21.7
	157	35.1	0.80	23.3
Leeward wall	All	35.1	−0.30	−8.7
Side walls	All	35.1	−0.70	−20.4
Roof	0–79	35.1	−0.98	−28.6
	79–157	35.1	−0.78	−22.7
	157–200	35.1	−0.62	−18.1

Note: q_h = 35.1 psf; G = 0.83.

For partially enclosed buildings:

$$GC_{pi} = \pm 0.55 \hspace{2cm} \text{(Table 6-7)}$$

For q_i, q_h = 35.1 psf for negative internal pressure, and q_z will be evaluated at 60 ft for positive internal pressure (the point at which the enclosed building classification changes to partially enclosed).

Internal Pressure Calculation

Negative internal pressure = 35.1 × (−0.55) = −19.3 psf

Positive internal pressure = 26.6 × 0.55 = 14.6 psf (q_z is gained by interpolation at height z = 60 ft)

Parapet Load on MWFRS

According to Section 6.5.12.2.4 of the Standard:

$$P_p = q_p GC_{pn} \hspace{2cm} \text{(Eq. 6-20)}$$
$$q_p = 35.4 \text{ psf}$$
$$GC_{pn} = 1.5 \text{ for windward parapet}$$
$$= -1.0 \text{ for leeward parapet}$$

Parapet is 3 ft high; the force on parapets of MWFRS can be determined as follows:

$$F = 35.4 \times 1.5 \times 3 = 159.3 \text{ plf for windward parapet}$$
$$= 35.4 \times (-1.0) \times 3 = -106.2 \text{ plf for leeward parapet}$$

This force is to be applied on the windward parapet and to the leeward parapet.

Design wind pressures for MWFRS are shown in **Figure G3–7** for wind normal to 200-ft face and in **Figure G3–8** for wind normal to 100-ft face.

For design of parapet, see the loads on components and cladding.

3.3.6 Design Wind Load Cases

Section 6.5.12.3 of the Standard requires that any building whose wind loads have been determined under the provisions of Sections 6.5.12.2.1 and 6.5.12.2.3 shall be designed for wind load cases as defined in Figure 6-9. Case 1 includes the loadings determined in this example and shown in **Figures G3–7** and **G3–8**. A combination of windward (P_W) and leeward (P_L) loads are applied for Load Cases 2, 3, and 4 as shown in **Figure G3–9**.

3.3.7 Design Pressures for C&C

Design pressure for C&C is obtained by the following equation:

$$p = q(GC_p) - q_i(GC_{pi}) \tag{Eq. 6-23}$$

where

q = q_z for windward wall calculated at height z and q_h for leeward wall, side walls, and roof calculated at height h

q_i = q_h = 35.1 psf for negative internal pressure

= q_z evaluated at 60 ft = 26.6 psf for positive internal pressure

(GC_p) = External pressure coefficient (see Figure 6-17 of the Standard)

(GC_{pi}) = Internal pressure coefficient (see Figure 6-5 of the Standard)

Wall Design Pressures

The pressure coefficients (GC_p) are a function of effective wind area (see **Table G3–8**). The definition of effective wind area for a C&C panel is the span length multiplied by an effective width that need not be less than one-third the span length (see Section 6.2 of the Standard). The effective wind areas, A, for wall components are:

Mullion

larger of $A = 11(5) = 55 \text{ ft}^2$ (controls)
or $A = 11(11/3) = 40.3 \text{ ft}^2$

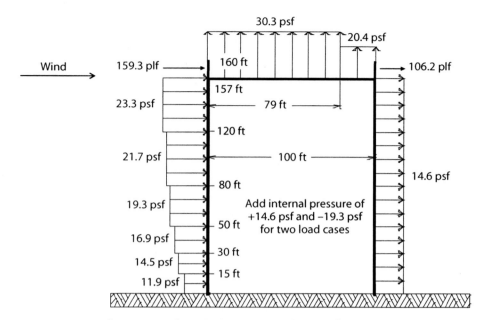

Figure G3–7 Design Pressures for MWFRS for Wind Normal to the 200-ft Face

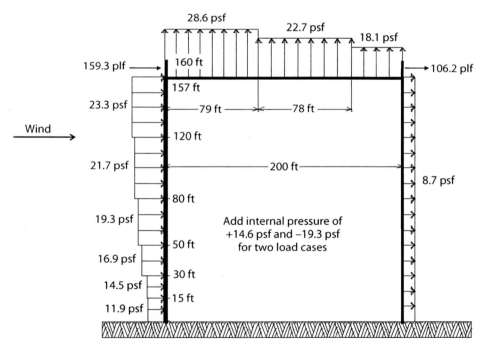

Figure G3–8 Design Pressures for MWFRS for Wind Normal to the 100-ft Face

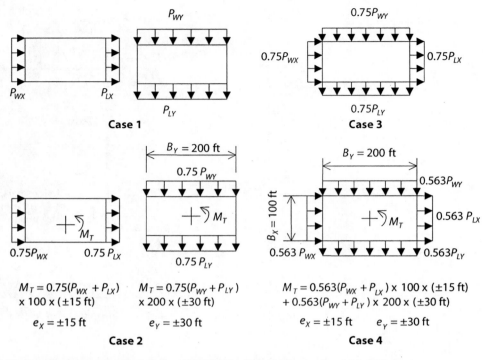

Figure G3–9 Design Pressures in Case B for MWFRS for Wind Normal to 100-ft Face

Glazing panel

    larger of $A = 5(5.5) = 27.5 \text{ ft}^2$ (controls)

    or $A = 5(5/3) = 8.3 \text{ ft}^2$

Width of corner Zone 5

    larger of $a = 0.1(100) = 10 \text{ ft}$ (controls)

    or $a = 3 \text{ ft}$

The internal pressure coefficient $(GC_{pi}) = \pm 0.55$ (Figure 6-5).

See notes above about the location of enclosed building area and internal pressure coefficient of $GC_{pi} = \pm 0.18$ can be used in the bottom 60 ft.

Table G3–8 Wall (GC_p) for Ex. 3

Component	A (ft^2)	Zones 4 and 5 $(+GC_p)$	Zone 4 $(-GC_p)$	Zone 5 $(-GC_p)$
Mullion	55	0.81	−0.84	−1.55
Panel	27.5	0.87	−0.88	−1.72

Typical Design Pressure Calculations

Design pressures for mullions are shown in **Table G3–9** and for panels in **Table G3–10**. Controlling negative design pressure for mullion in Zone 4 of walls for $h = 60$ ft and above:

$$=35.1(-0.84) - 26.6 \times 0.55$$

$$=-44.1 \text{ psf (positive internal pressure controls)}$$

Controlling negative design pressure for mullion in Zone 4 for wall below 60 ft:

$$=35.1(-0.84) - 26.6 \times 0.18$$

$$= -34.3 \text{ psf (positive internal pressure controls).}$$

Table G3–9   Controlling Design Pressures for Mullions (psf)

	Design Pressure			
	Zone 4		Zone 5	
z (ft)	Positive	Negative	Positive	Negative
0–15	37.0	–34.3	37.0	–59.2
15–30	37.0	–34.3	37.0	–59.2
30–50	39.9	–34.3	39.9	–59.2
50–80	42.9	–44.1	42.9	–69.0
80–120	45.7	–44.1	45.7	–69.0
120–157	47.7	–44.1	47.7	–69.0

Table G3–10   Design Pressures of Panels (psf)

	Design Pressure			
	Zone 4		Zone 5	
z (ft)	Positive	Negative	Positive	Negative
0–15	38.4	–45.5	38.4	–75.0
15–30	38.4	–45.5	38.4	–75.0
30–50	41.4	–45.5	41.4	–75.0
50–80	44.6	–45.5	44.6	–75.0
80–120	47.7	–45.5	47.7	–75.0
120–157	49.8	–45.5	49.8	–75.0

Controlling positive design pressure for mullion in Zone 4 of walls at roof height:

$$= 35.1 \times 0.81 - 35.1 \times (-0.55)$$

$$= 47.7 \text{ psf (negative internal pressure controls)}$$

Controlling negative pressure is obtained with positive internal pressure, and controlling positive pressure is obtained with negative internal pressure.

Parapet Design Pressures

The design wind pressure on the C&C elements of parapets shall be determined according to the following equation (Section 6.5.12.4.4 of the Standard). In this example, the effective wind area is assumed to be 3 ft × 3 ft = 9 ft².

$$p = q_p (GC_p - GC_{pi}) \qquad \text{(Eq. 6-24)}$$

where

q_p = Velocity pressure evaluated at the top of parapet.

GC_p = External pressure coefficient from Figures 6-11 through 6-17 of the Standard.

GC_{pi} = Internal pressure coefficient from Figure 6-5 of the Standard, based on the porosity of the parapet envelope. In this example, internal pressure is not included since parapet is assumed to be nonporous.

Note that, according to Note 7 of Figure 6-17, Zone 3 is treated as Zone 2.

Load Case A

$35.4 \times [(0.9) - (-2.3)] = 113.3$ psf (directed inward)

Load Case B

$35.4 \times [(0.9) - (-1.8)] = 95.6$ psf (directed outward)

Roof Design Pressures

The C&C roof pressure coefficients are given in Figure 6-17 of the Standard. The pressure coefficients (**Table G3–11**) are a function of the effective wind area. Since specific components of roofs are not identified, design pressures are given for various effective wind areas, A.

The design pressures are the algebraic sum of external and internal pressures. Positive internal pressure provides controlling negative pressures. These design pressures act across the roof surface (interior to exterior):

Design internal pressures = $26.6 \times 0.55 = 14.6$ psf

Design pressures = $q_h (GC_p) - 14.6 = 35.1 (GC_p) - 14.6$

Design pressures are summarized in **Table G3–12**.

Table G3-11   Roof External Pressure Coefficient (GC_p)

A (ft^2)	Zone 1 GC_p	Zones 2 and 3 $-GC_p$
≤ 10	−1.40	−2.30
20	−1.31	−2.18
100	−1.11	−1.89
250	−0.99	−1.72
400	−0.93	−1.64
≥ 500	−0.90	−1.60

Note: Note 7 in Figure 6–17 of the Standard permits treatment of Zone 3 as Zone 2 if parapet of 3 ft or higher is provided.

Table G3-12   Roof Design Pressures (psf)

$A(ft^2)$	Design pressures negative	
	Zone 1	Zones 2 and 3
≤ 10	−63.7	−95.3
20	−60.6	−91.1
100	−53.6	−80.9
250	−49.3	−75.0
400	−47.2	−72.2
500	−46.2	−70.8

3.4 Example 4: Office Building from Ex. 3 Located on an Escarpment

In this example, velocity pressures for the office building of Ex. 3, when it is located on an escarpment, are determined. Design pressures for MWFRS and C&C can be determined in the same manner as Ex. 3 once velocity pressures q_z and q_h are determined. The building is illustrated in **Figure G3–10**; data are provided below.

Location   City in Alaska

Topography   Escarpment as shown

Terrain   Suburban

Dimensions 100 ft × 200 ft in plan
Roof height of 157 ft with 3-ft parapet
Flat roof

Framing Reinforced concrete rigid frame in both directions
Floor and roof slabs provide diaphragm action
Fundamental natural frequency is greater than 1 Hz

Cladding Mullions for glazing panels span 11 ft between floor slabs
Mullion spacing is 5 ft

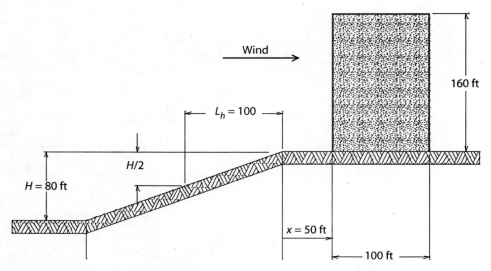

Figure G3–10 Building Characteristics for Example 4, Office Building on Escarpment. Notes: 1) L_h is measured from mid-height to top of the slope. 2) x distance is taken to the front of the building as a conservative value.

Glazing panels are 5-ft wide × 5-ft 6 in. high (typical). Glazing does not have to be wind-borne debris impact resistant because Alaska is not in a hurricane-prone region (see Section 6.5.9.3 of the Standard).

3.4.1 Exposure, Building Classification, and Basic Wind Speed

Exposure B, same as Ex. 3

Category II

$V = 120$ mph, same as Ex. 3

3.4.2 Velocity Pressures

The velocity pressure equation is

$$q_z = 0.00256 K_z K_{zt} K_d V^2 I \text{ psf}$$

where

K_z = value obtained from Table 6-3 of the Standard

K_{zt} = value determined using Figure 6-4 of the Standard

K_d = 0.85 for buildings (see Table 6-4 of the Standard)

V = 120 mph

I = 1.0 for Category II from Table 6-1

Determination of K_{zt}

The topographic effect of escarpment applies only when the upwind terrain is free of topographic features for a distance equal to 100 H or 2 mi, whichever is smaller. For this example, it is assumed that there are no topographic features upwind for a distance of 8,000 ft.

For use in Figure 6-4 of the Standard,

H = 80 ft

L_h = 100 ft

x = 50 ft (distance to the front face of the building)

Since $H/L_h = 0.8 > 0.5$, according to Note 2 in Figure 6-4 of the Standard, use $H/L_h = 0.5$ and $L_h = 2H = 160$ ft.

The building is on a 2-D escarpment.

For Exposure B

$K_1/(H/L_h) = 0.75$, therefore $K_1 = (0.75)(0.5) = 0.38$ (Figure 6-4)

For x/L_h = 50 ft/160 ft = 0.31; $K_2 = 1 - (0.31/4) = 0.92$ (Figure 6-4)

$$K_3 = e^{-2.5z/L_h} \text{ (values in table for } z\text{)}$$

$$K_{zt} = (1 + K_1 K_2 K_3)^2 \quad \text{(Eq. 6-3)}$$

$$q_z = 0.00256 K_z K_{zt}(0.85)(120)^2(1.0)$$

Values for q_z are shown in **Table G3–13**.

3.4.3 Effect of Escarpment

Velocity pressures q_z are compared with the values of Ex. 3 in **Table G3–14** to assess the effect of the escarpment. The increase in velocity pressures does not directly translate into an increase in design pressures as discussed below.

For MWFRS, the external windward wall pressures will increase by the percentages shown at various heights; however, the external leeward wall, side wall, and roof pressures will increase by 8% since these pressures are

Table G3–13 Speed-up Velocity Pressures (psf)

Height (ft)	K_z	z/L_h	K_3	K_{zt}	q_z (psf)
0–15	0.57	0.05	0.88	1.71	30.5
30	0.70	0.14	0.71	1.56	34.2
50	0.81	0.25	0.54	1.41	35.8
80	0.93	0.41	0.36	1.27	37.0
120	1.04	0.63	0.21	1.15	37.5
h = 157	K_h = 1.12	0.87	0.11	1.08	37.9

Notes: z is taken midway between the height range because it is unconservative for K_{zt} to take top height of the range. L_h = 160 ft.

Table G3–14 Velocity Pressure q_z (psf)

Height (ft)	Homogeneous terrain (Ex. 3)	Escarpment (Ex. 4)	Percentage increase
0–15	17.9	30.5	71
30	21.9	34.2	56
50	25.4	35.8	41
80	29.1	37.0	27
120	32.6	37.5	15
157 (roof)	35.1	37.9	8

controlled by velocity pressure at roof height, q_h. Internal pressures will depend on assessment of openings.

For C&C, the negative (outward acting) external pressures will also increase by only 8%.

3.5 Example 5: 2,500-ft² House with Gable/Hip Roof

Design wind pressures for a typical one-story house are to be determined. Various views of the house are provided in **Figures G3–11(a)–(d)**. The physical data are as shown here.

Location Dallas–Fort Worth, Texas

Topography Homogenous

Terrain Suburban

Dimensions 80 ft × 40 ft (including porch) footprint
Porch is 8 ft × 48 ft
Wall eave height is 10 ft
Roof gable Θ = 15°; roof overhang is 2 ft all around

Framing Typical timber construction
Wall studs are spaced 16 in. on center
Roof trusses spanning 32 ft are spaced 2 ft on center
Roof panels are 4 ft × 8 ft

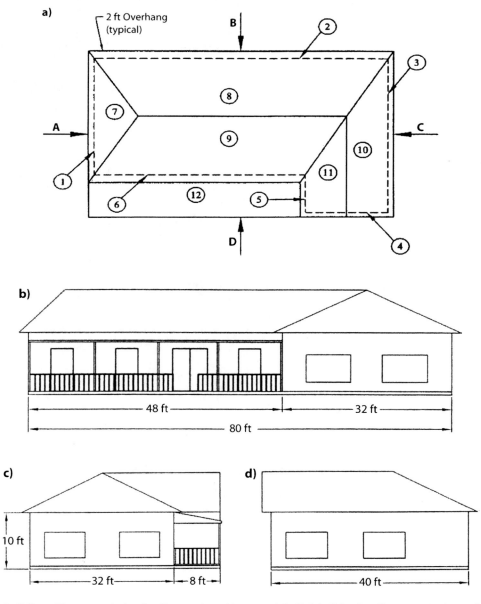

Figure G3–11a–d Building Characteristics for Example 5, House with Gable/Hip Roof. Note: a) View of roof b) view of front c) view of Side A, and d) view of Side C

Glazing is uniformly distributed (pressures on C&C will depend on effective area and location; for brevity, all items are not included).

Wind speed $V = 90$ mph

Importance factor $I = 1.0$

Topography factor $K_{zt} = 1.0$

Directionality factor $K_d = 0.85$ (for buildings)

The building is located in a suburban area; according to Section 6.5.6.2 and 6.5.6.3 of the Standard, Exposure B is used.

$$\text{mean roof height} = 10 + \frac{(16)(\tan 15°)}{2} = 12.1 \text{ ft}$$

Since K_z is constant in the 0 to 15 ft region, from Table 6-3 of the Standard,

$K_z = K_h = 0.70$ for Case 1 (C&C)

$K_z = K_h = 0.57$ for Case 2 (MWFRS)

3.5.1 Velocity Pressures

$q_z = 0.00256\, K_z K_{zt} K_d\, V^2\, I$ psf (Eq. 6-15)

For MWFRS

$q_z = q_h = 0.00256\, (0.57)\, (1.0)\, (0.85)\, (90)^2\, (1.0) = 10.0$ psf

For C&C

$q_z = q_h = 0.00256\, (0.7)\, (1.0)\, (0.85)\, (90)^2\, (1.0) = 12.3$ psf

Gust Effect Factor

$G = 0.85$ (Section 6.5.8.1)

$(GC_{pi}) = +0.18$ and -0.18 (Figure 6-5)

3.5.2 Wind Pressure for MWFRS

Because of asymmetry, all four wind directions are considered (normal to walls). The wall surfaces are numbered 1 through 6; roof surfaces are 7 through 11; porch roof surface is 12. The external pressure coefficients are from Figure 6-6.

Wind Direction A

Wall pressures

Surface 1: $p = 10.0\ (0.85)\ (0.8) - 10.0\ (\pm 0.18) = +6.8 \pm 1.8$ psf (windward)

Surface 2: $p = 10.0\ (0.85)\ (-0.7) - 10.0\ (\pm 0.18) = -6.0 \pm 1.8$ psf (side)

Surface 3: $p = 10.0\ (0.85)\ (-0.3) - 10.0\ (\pm 0.18) = -2.6 \pm 1.8$ psf (leeward)
(for $L/B = 80/40 = 2$; $C_p = -0.3$)

Surface 4: $p = -6.0 \pm 1.8$ psf (side)

Surface 5: $p = +6.8 \pm 1.8$ psf (windward)

Surface 6: $p = -6.0 \pm 1.8$ psf (side)

Roof pressures

Roof C_p from Figure 6-3 of the Standard is shown in **Table G3–15**.

Roof pressures calculation

Surface 7: $p = 10.0\ (0.85)\ (-0.5) - 10.0\ (\pm 0.18) = -4.2 \pm 1.8$ psf (windward)

Surface 8: for $\Theta = 0°$; pressure varies along the roof

$p = 10.0\ (0.85)\ (-0.9) - 10.0\ (\pm 0.18) = -7.6 \pm 1.8$ psf; 1 to 12.1 ft
$p = 10.0\ (0.85)\ (-0.5) - 10.0\ (\pm 0.18) = -4.2 \pm 1.8$ psf; 12.1 to 24.2 ft
$p = 10.0\ (0.85)\ (-0.3) - 10.0\ (\pm 0.18) = -2.6 \pm 1.8$ psf; 24.2 ft to end

Surface 9: Same pressures as surface 8

Surface 10: $p = 10.0\ (0.85)\ (-0.5) - 10.0\ (\pm 0.18) = -4.2 \pm 1.8$ psf (leeward)

Surface 11: $p = 10.0\ (0.85)\ (-0.5) - 10.0\ (\pm 0.18) = -4.2 \pm 1.8$ psf (windward)

Surface 12: Same as surface 8 without internal pressure

Table G3–15   Roof C_p^* for Wind Direction A, by Surface

Surface	7	8			9	10	11	12
C_p	−0.5	*Horiz. distance from windward edge (ft)*			Same as surface 8	−0.5	−0.5	Same as surface 8
	0.0*	1 to 12.1	12.1 to 24.2	24.2 to end		−0.5*	0.0*	
		−0.9	−0.5	−0.3				
		−0.18*	−0.18*	−0.18*				

* The values of smaller uplift pressures (lower row) on the roof can become critical when wind load is combined with roof live load or snow load; load combinations are given in Sections 2.3 and 2.4 of the Standard. For brevity, loading for this value is not shown here.

Overhang pressures

At wall surfaces 1 and 5,

$$p = 10.0 \, (0.85) \, (0.8) = +6.8 \text{ psf}$$

Internal pressure is of the same sign on all applicable surfaces.

Wind Direction B

Wall pressures

Surface 1: $p = -6.0 \pm 1.8$ psf (side)

Surface 2: $p = +6.8 \pm 1.8$ psf (windward)

Surface 3: $p = -6.0 \pm 1.8$ psf (side)

Surface 4: $p = 10.0 \, (0.85) \, (-0.5) - 10.0 \, (\pm 0.18) = -4.2 \pm 1.8$ psf (leeward)
(for $L/B = 40/80 = 0.5$; $C_p = -0.5$)

Surface 5: Even though technically this surface is side wall, it is likely to see the same pressure as surface 6

Surface 6: Same pressure as surface 4

Roof pressures

$h/L = 12.1/40 = 0.3$; $\Theta = 15°$

Roof C_p from Figure 6-6 of the Standard are tabulated in **Table G3–16**.

For windward, $C_p = -0.54$ (interpolated between -0.5 and -0.7)

For leeward, $C_p = -0.5$

For parallel to ridge

$C_p = -0.9$; 1 to 12.1 ft
$C_p = -0.5$; 12.1 to 24.2 ft
$C_p = -0.3$; 24.2 ft to end

Roof pressures calculation

Surface 7: Same pressures as surface 8 for Wind Direction A

Surface 8: $p = 10.0 \, (0.85) \, (-0.54) - 10.0 \, (\pm 0.18) = -4.6 \pm 1.8$ psf (windward)

Surface 9: $p = 10.0 \, (0.85) \, (-0.5) - 10.0 \, (\pm 0.18) = -4.2 \pm 1.8$ psf (leeward)

Table G3–16 Roof C_p for Wind Direction B, by Surface

7	8	9	10	11	12
Same as surface 8 for Wind Direction A	-0.54*	-0.5	Same as surface 8 for Wind Direction A	Same as surface 9	-0.3

* Value is gained by interpolation

Surface 10: Same pressures as surface 8 for Wind Direction A

Surface 11: Same as surface 9 because it is sloping with respect to ridge

Surface 12: This surface is at a distance greater than $2h$

$p = 10.0\ (0.85)\ (-0.3) = -2.6$ psf; no internal pressure

Overhang pressures at wall surface 2,

$p = 10.0\ (0.85)\ (0.8) = +6.8$ psf

Internal pressure is of the same sign on all applicable surfaces.

Wind Direction C

Wall pressures

Surfaces 1 and 5: $p = -2.6 \pm 1.8$ psf (leeward)

Surfaces 2, 4, and 6: $p = -6.0 \pm 1.8$ psf (side)

Surface 3: $p = +6.8 \pm 1.8$ psf (windward)

Roof pressure

Surfaces 7 and 11: $p = -4.2 \pm 1.8$ psf (leeward)

Surfaces 8 and 9: Pressures vary along the roof; same pressures as surface 8 for Wind Direction A

Surface 10: $p = -4.2 \pm 1.8$ psf (windward)

Surface 12: Same pressures as surface 9 without internal pressures

Overhang pressures at wall surface 3

$p = 10.1\ (0.85)\ (0.8) = +6.8$ psf

Internal pressure is of the same sign on all applicable surfaces.

Wind Direction D

Wall pressures

Surfaces 1 and 3: $p = -6.0 \pm 1.8$ psf (side)

Surface 2: $p = -4.2 \pm 1.8$ psf (leeward)

Surfaces 4, 5, and 6: $p = +6.8 \pm 1.8$ psf (windward)

Roof pressures

Surfaces 7, 10, and 11: Pressures vary along the roof; same pressures as surface 8 for Wind Direction A

Surface 8: $p = -4.2 \pm 1.8$ psf (leeward)

Surface 9: $p = -4.2 \pm 1.8$ psf (windward)

Surface 12: This surface will see pressures on top and bottom surfaces; they will add algebraically.

For $\Theta = 0°$, $h/L < 0.5$, $C_p = -0.9$ (Load Case 1)

$p = 10.0\ (0.85)\ (-0.9) - 10.0\ (0.85)\ (+0.8) = -14.4$ psf uplift

Overhang pressures at wall surfaces 4, 5, and 6,

$p = 10.0\ (0.85)\ (0.8) = +6.8$ psf

Internal pressure is of the same sign on all applicable surfaces.

3.5.3 Design Wind Load Cases

Section 6.5.12.3 of the Standard requires that any building whose wind loads have been determined under the provisions of Sections 6.5.12.2.1 and 6.5.12.2.3 shall be designed for wind load cases as defined in Fig. 6-9. Case 1 includes the loadings analyzed above. A combination of windward (P_W) and leeward (P_L) loads is applied for other Load Cases. This building has mean roof height h of less than 30 ft; hence it comes under exception specified in Section 6.5.12.3 of the Standard. Only Load Cases 1 and 3 shown in Figure 6-9 of the Standard have to be considered.

The following two points need to be highlighted:

1. Because of asymmetry, all four wind directions are considered when combining wind loads according to Figure 6-9 of the Standard. For example, when combining wind loads in Case 3, there are four kinds of combinations of wind loads that need to be considered, which are shown as in **Figure G3–12**.
2. Due to the slightness of the roof slope, the wind load acting on the roof is negligible here.

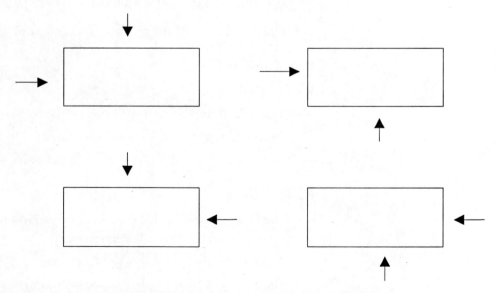

Figure G3–12   Combinations of Wind Loads. (Note: Arrows show the wind direction.)

3.5.4 Design Pressures for C&C

Wall Component

Wall studs are 10-ft long and spaced 16 in. apart.

Effective area

> larger of $\quad\quad 10 \times 1.33 = 13.3$ ft^2
> or $\quad\quad\quad\quad 10 \times 10/3 = 33.3$ ft^2 (controls)

From Figure 6-11A of the Standard, equations in Chapter 2 of this guide are used:

> $(GC_p) = +0.91$ for Zones 4 and 5
> $(GC_p) = -1.01$ for Zone 4
> $(GC_p) = -1.22$ for Zone 5

Distance "a"

> smaller of $\quad\quad 0.1 (40) = 4$ ft (controls)
> or $\quad\quad\quad\quad 0.4 (12.1) = 4.8$ ft

Design pressure

> $p = 12.3 \ (0.91 + 0.18) = +13.4$ psf (all walls)
> $p = 12.3 \ (-1.01 - 0.18) = -14.6$ psf (middle)
> $p = 12.3 \ (-1.22 - 0.18) = -17.2$ psf (corner)

Roof Component

Distance "a"

> smaller of $\quad\quad 0.1 (40) = 4$ ft (controls)
> or $\quad\quad\quad\quad 0.4 (12.1) = 4.8$ ft

Roof trusses are 32 ft long and spaced 2 ft apart.

Effective area

> larger of $\quad\quad 32 \times 2 = 64$ ft^2
> or $\quad\quad\quad\quad 32 \times 32/3 = 341$ ft^2 (controls)

From Figure 6-11C of the Standard, for $\Theta = 15°$,

> $(GC_p) = +0.3$ for Zones 1, 2, and 3 (Note: Zone 3 covers very small area of truss)
> $(GC_p) = -0.8$ for Zone 1
> $(GC_p) = -2.2$ for Zone 2 (includes overhang area)

Design pressures

$p = 12.3\ (0.3 + 0.18) = +5.9$ psf (all zones)

$p = 12.3\ (-0.8 - 0.18) = -12.1$ psf (Zone 3)

$p = 12.3 \times (-2.2) = -27.1$ psf (Zone 2)

Overhang pressures to be used for reaction and anchorage

$p = 12.3\ (-2.2) = -27.1$ psf (edge of roof)

$p = 12.3\ (-2.5) = -30.8$ psf (roof corners)

(MWFRS pressures should be used for the truss connection loads across the building span; C&C loads are used for the additional pressure created by the presence of the overhang.)

Roof Panels

Effective area = $4 \times 8 = 32$ ft^2. From Figure 6-11C of the Standard, for $\Theta = 15°$ (Note: Zones 2 and 3 are regarded as overhang),

$(GC_p) = +0.4$ for Zones 1, 2, and 3

$(GC_p) = -0.85$ for Zone 1

$(GC_p) = -2.2$ for Zone 2 (with overhang)

$(GC_p) = -2.2$ for Zone 3 on hip roofs (with overhang)

$(GC_p) = -3.1$ for Zone 3 on gable roofs (with overhang)

Design pressures

$p = 12.3\ (0.4 + 0.18) = +7.1$ psf (all zones)

$p = 12.3 \times (-0.85 - 0.18) = -12.7$ psf (Zones 1)

$p = 12.3 \times (-2.2) = -27.1$ psf (Zone 2)

$p = 12.3 \times (-2.2) = -27.1$ psf (Zone 3 on hip roofs)

$p = 12.3 \times (-3.1) = -38.1$ psf (Zone 3 on gable roofs)

Fasteners

Effective area = 10 ft^2

$(GC_p) = +0.5$ for Zones 1, 2, and 3

$(GC_p) = -0.9$ for Zone 1

$(GC_p) = -2.2$ for Zones 2 (with overhang)

$(GC_p) = -2.2$ for Zone 3 on hip roofs (with overhang)

$(GC_p) = -3.7$ for Zone 3 on gable roofs (with overhang)

Design pressures

$p = 12.3 (0.5 + 0.18) = +8.4$ psf (all zones)

$p = 12.3 \times (-0.9 - 0.18) = -13.3$ psf (Zones 1)

$p = 12.3 \times (-2.2) = -27.1$ psf (Zone 2)

$p = 12.3 \times (-2.2) = -27.1$ psf (Zone 3 on hip roofs)

$p = 12.3 \times (-3.7) = -45.5$ psf (Zone 3 on gable roofs)

3.6 Example 6: 200-ft × 250-ft Gable Roof Commercial/Warehouse Building Using Buildings of All Height Provisions

In this example, design wind pressures for a large, one-story commercial/warehouse building are determined. **Figure G3–13** shows the dimensions and framing of the building. The building data are as shown here.

Location Memphis, Tennessee

Terrain Flat farmland

Dimensions 200 ft × 250 ft in plan
Eave height of 20 ft
Roof slope 4:12 (18.4°)

Framing Rigid frames span the 200-ft direction
Rigid frame bay spacing is 25 ft
Lateral bracing in the 250-ft direction is provided by a "wind truss" spanning the 200 ft to side walls and cable/rod bracing in the planes of the walls
Girts and purlins span between rigid frames (25-ft span)
Girt spacing is 6 ft 8 in.
Purlin spacing is 5 ft

Cladding Roof panel dimensions are 2 ft wide
Roof fastener spacing on purlins is 1 ft on center
Wall panel dimensions are 2 ft × 20 ft
Wall fastener spacing on girts is 1 ft on center
Openings are uniformly distributed

3.6.1 Exposure and Building Classification

The building is located on flat and open farmland. It does not fit Exposures B or D; therefore, Exposure C is used (Sections 6.5.6.2 and 6.5.6.3 of the Standard).

The building function is commercial-industrial. It is not considered an essential facility or likely to be occupied by 300 persons at one time. Category II is appropriate. Table 1-1 and Table 6-1 of the Standard specify an importance factor $I = 1.0$.

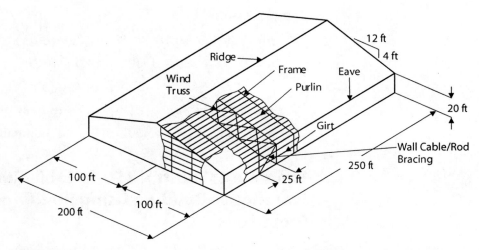

Figure G3–13 Building Characteristics for Examples 6 and 7, Commercial/Warehouse Building

3.6.2 Basic Wind Speed

Selection of basic wind speed is addressed in Section 6.5.4 of the Standard. Memphis, Tennessee, is not located in the special wind region, nor is there any reason to suggest that winds at the site are unusual and require additional attention. Therefore, the basic wind speed is $V = 90$ mph (see Figure 6-1 of the Standard).

3.6.3 Analytical Procedure

Method 2, Analytical Procedure, is used in this example (see Section 6.5 of the Standard). In addition, provisions of buildings of all heights, given in Section 6.5.12.2.1 and Figure 6-6 for MWFRS, will be used. Alternate provisions of low-rise buildings are illustrated in Ex. 7 (Section 3.7 of this guide).

3.6.4 Wind Directionality

Wind directionality factor is given in Table 6-4 of the Standard. For MWFRS and C&C, the factor $K_d = 0.85$.

3.6.5 Velocity Pressures

The velocity pressures (**Table G3–17**) are computed using

$$q_z = 0.00256\ K_z K_{zt} K_d V^2 I\ \text{psf} \qquad \text{(Eq. 6-15)}$$

where

K_z = Value obtained from Table 6-3 of the Standard

K_{zt} = 1.0 (no topographic effect)

I = 1.0 for Category II building

K_d = 0.85

V = 90 mph

Table G3–17 External Wall C_p

Height	ft	K_z	q_z (psf)
	0–15	0.85	15.0
Eave	20	0.90	15.8
	30	0.98	17.3
h	36.7	1.02	18.0*
	40	1.04	18.3
	50	1.09	19.2
Ridge	53.3	1.10	19.4

* q_h = 18.0 psf.

therefore,

$$q_z = 0.00256 K_z (1.0)(0.85)(90)^2 (1.0) = 17.6\, K_z \text{ psf}$$

Values for K_z are the same for Cases 1 and 2 for Exposure C (see Table 6-3 of the Standard). Mean roof height $h = 36.7$ ft.

3.6.6 Design Wind Pressure

Design wind pressures for MWFRS of this building can be obtained using Section 6.5.12.2.1 of the Standard for buildings of all heights or Section 6.5.12.2.2 for low-rise buildings. Pressures determined in this example are using buildings of all heights criteria. Ex. 7 illustrates use of low-rise building criteria.

$$p = qGC_p - q_i(GC_{pi}) \tag{Eq. 6-17}$$

where

$q = q_z$ for windward wall at height z above ground

$q = q_h$ for leeward wall, side walls, and roof

$q_i = q_h$ for enclosed buildings

G = Gust effect factor

C_p = Values obtained from Figure 6-6 of the Standard

(GC_{pi}) = Values obtained from Figure 6-5

For this example, when the wind is normal to the ridge, the windward roof experiences both positive and negative external pressures. Combining these external pressures with positive and negative internal pressures will result in four loading cases when wind is normal to the ridge.

When wind is parallel to the ridge, positive and negative internal pressures result in two loading cases. The external pressure coefficients, C_p for $\theta = 0°$, apply in this case.

Gust Effect Factor

For rigid structures, G can be calculated using Eq. 6-4 (see Section 6.5.8.1 of the Standard) or alternatively taken as 0.85. For simplicity, $G = 0.85$ is used in this example.

External Wall Pressure

The pressure coefficients for the windward wall and for the side walls (see Figure 6-6 of the Standard) are 0.8 and –0.7, respectively, for all L/B ratios.

The leeward wall pressure coefficient is a function of the L/B ratio. For wind normal to the ridge, $L/B = 200/250 = 0.8$; therefore, the leeward wall pressure coefficient is –0.5. For flow parallel to the ridge, $L/B = 250/200 = 1.25$; the value of C_p is obtained by linear interpolation. The wall pressure coefficients are summarized in **Table G3–18**.

External Roof Pressure for Wind Normal to Ridge

The roof pressure coefficients for the MWFRS (**Table G3–19**) are obtained from Figure 6-6 of the Standard. For the roof angle of 18.4°, linear interpolation is used to establish C_p. For wind normal to the ridge, $h/L = 36.7/200 = 0.18$; hence, only single linear interpolation is required. Note that interpolation is only carried out between values of the same sign.

Internal (GC_{pi})

Values for (GC_{pi}) for buildings are addressed in Section 6.5.11.1 and Figure 6-5 of the Standard. The openings are evenly distributed in the walls (enclosed

Table G3–18 External Wall C_p

Surface	Wind direction	L/B	C_p
Windward wall	All	All	0.80
Leeward wall	Normal to ridge	0.8	–0.50
	Parallel to ridge	1.25	–0.45*
Side wall	All	All	–0.70

* By linear interpolation.

Table G3–19 Roof C_p (Wind Normal to Ridge)

Surface	15°	18.4°	20°
Windward roof	–0.5	–0.36*	–0.3
	0.0	0.14*	0.2
Leeward roof	–0.5	–0.57*	–0.6

* By linear interpolation.

building) and Memphis, Tennessee, is not in a hurricane-prone region. The reduction factor of Section 6.5.11.1.1 is not applicable for enclosed buildings; therefore,

$(GC_{pi}) = \pm 0.18$

MWFRS Net Pressures

$$p = qGC_p - q_i(GC_{pi}) \qquad \text{(Eq. 6-17)}$$

$$p = q(0.85)C_p - 18.0(\pm 0.18)$$

where

$q = q_z$ for windward wall

$q = q_h$ for leeward wall, side wall, and roof

$q_i = q_h$ for windward walls, side walls, leeward walls, and roofs of enclosed buildings

Typical Calculation

Windward wall, 0–15 ft, wind normal to ridge

$p = 15.0(0.85)(0.8) - 18.0(\pm 0.18)$

$p = 7.0$ psf with (+) internal pressure

$p = 13.4$ psf with (−) internal pressure

The net pressures for the MWFRS are summarized in **Table G3–20**.

External Roof Pressures for Wind Parallel to Ridge

For wind parallel to the ridge, $h/L = 36.7/250 = 0.147$ and $\theta < 10°$. The values of C_p for wind parallel to ridge are obtained from Figure 6-6 of the Standard and are shown in **Tables G3–21** and **G3–22.**

3.6.7 Design Wind Load Cases

Section 6.5.12.3 of the Standard requires that any building whose wind loads have been determined under the provisions of Sections 6.5.12.2.1 and 6.5.12.2.3 shall be designed for wind load cases as defined in Figure 6-9 of the Standard. Case 1 includes the loadings shown in **Figures G3–14** through **G3–17**. A combination of windward (P_W) and leeward (P_L) loads is applied for Load Cases 2, 3, and 4 as shown in Figure 6-9 of the Standard. Section 6.5.12.3 of the Standard has an exception that if a building is designed with flexible diaphragm, only Load Cases 1 and 3 need to be considered. There is not enough structural information given in this example to assess flexibility of roof diaphragm. Structural designer will have to make a judgment for each building.

Table G3–20 MWFRS Pressures: Wind Normal to Ridge

Surface	z (ft)	q (psf)	G	C_p	Net pressure psf with $(+GC_{pi})$	Net pressure psf with $(-GC_{pi})$
Windward wall	0–15	15.0	0.85	0.8	7.0	13.4
	20	15.8	0.85	0.8	7.5	14.0
Leeward wall	All	18.0	0.85	–0.5	–10.9	–4.4
Side walls	All	18.0	0.85	–0.7	–14.0	–7.5
Windward Roof*	—	18.0	0.85	–0.36	–8.7	–2.3
				0.14	–1.1	5.4
Leeward roof	—	18.0	0.85	–0.57	–12.0	–5.5

Notes: q_h = 18.0 psf; (GC_{pi}) = ±0.18; $q_h(GC_{pi})$ = ±3.2 psf.

* Two loadings on windward roof and two internal pressures yield a total of four loading cases (see **Figures G3–14** and **G3–15**).

Table G3–21 Roof C_p (Wind Parallel to Ridge)

Surface	h/L	Distance from windward edge	C_p
Roof	≤ 0.5	0 to h	–0.9, –0.18*
		h to $2h$	–0.5, –0.18*
		> $2h$	–0.3, –0.18*

* The values of smaller uplift pressures on the roof can become critical when wind load is combined with roof live load or snow load; load combinations are given in Sections 2.3 and 2.4 of the Standard. For brevity, loading for this value is not shown in this example.

3.6.8 Design Pressures for C&C

Eq. 6-22 from Section 6.5.12.4 of the Standard is used to obtain the design pressures for components and cladding.

$$p = q_h[(GC_p) - (GC_{pi})] \quad \text{(Eq. 6-22)}$$

where

q_h = 18.0 psf

(GC_p) = Values obtained from Figure 6-11

(GC_{pi}) = ± 0.18 for this building

Table G3–22   MWFRS Pressures: Wind Parallel to Ridge

Surface	z (ft)	q (psf)	G	C_p	Net pressure psf with $(+GC_{pi})$	Net pressure psf with $(-GC_{pi})$
Windward wall	0–15	15.0	0.85	0.8	7.0	13.4
	20	15.8	0.85	0.8	7.5	14.0
	30	17.3	0.85	0.8	8.5	15.0
	40	18.3	0.85	0.8	9.2	15.7
	53.3	19.4	0.85	0.8	10.0	16.4
Leeward wall	All	18.0	0.85	−0.45	−10.1	−3.6
Side walls	All	18.0	0.85	−0.7	−14.0	−7.5
Roof	0 to h	18.0	0.85	−0.9	−17.0	−10.5
	h to $2h$	18.0	0.85	−0.5	−10.9	−4.4
	$> 2h$	18.0	0.85	−0.3	−7.8	−1.4

Notes: $q_h = 18.0$ psf; $(GC_{pi}) = \pm 0.18$; $h = 36.7$ ft; $q_h(GC_{pi}) = \pm 3.2$ psf. Roof measures distance from windward edge.

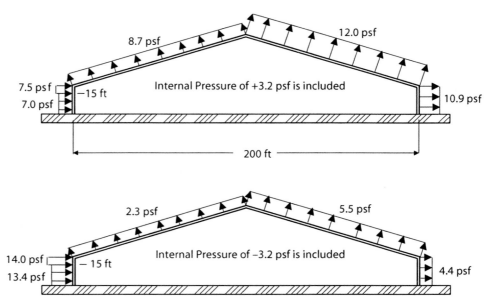

Figure G3–14   Net Design Pressures for MWFRS When Wind Is Normal to Ridge with Negative Windward External Roof Pressure Coefficient

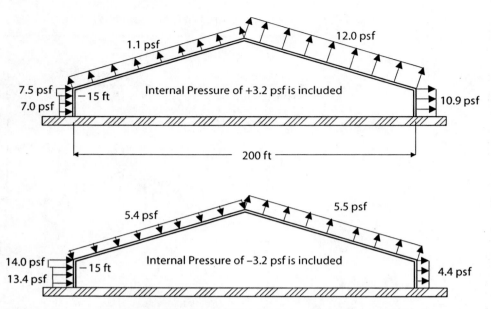

Figure G3–15   Net Design Pressures for MWFRS When Wind Is Normal to Ridge with Positive Windward External Roof Pressure Coefficient

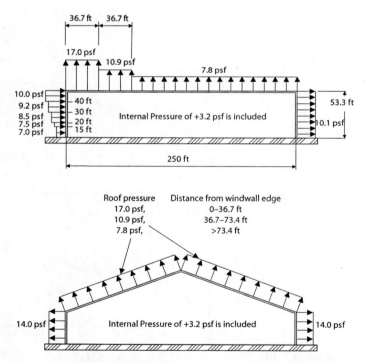

Figure G3–16   Net Design Pressures for MWFRS When Wind Is Parallel to Ridge with Positive Internal Pressure

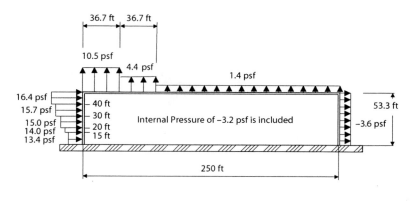

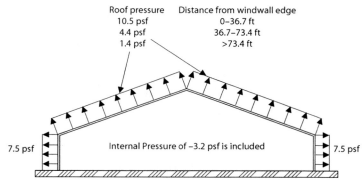

Figure G3–17 Net Design Pressures for MWFRS When Wind is Parallel to Ridge with Negative Internal Pressure

Wall Pressures

The C&C pressure coefficients (GC_p) (**Table G3–23**) are a function of effective wind area. The definitions of effective wind area for a component or cladding panel is the span length multiplied by an effective width that need not be less than one-third the span length; however, for a fastener it is the area tributary to an individual fastener.

Girt

> larger of $\quad A = 25(6.67) = 167 \text{ ft}^2$
> or $\quad\quad\quad A = 25(25/3) = 208 \text{ ft}^2$ (controls)

Wall Panel

> larger of $\quad A = 6.67(2) = 13.3 \text{ ft}^2$
> or $\quad\quad\quad A = 6.67(6.67/3) = 14.8 \text{ ft}^2$ (controls)

Fastener

> $A = 6.67(1) = 6.7 \text{ ft}^2$

Table G3–23 Wall Coefficients (GC_p), by Zone

		External (GC_p)		
C&C	$A(ft^2)$	Zones 4 and 5	Zone 4	Zone 5
Girt	208	0.77*	−0.87	−0.93
Panel	14.8	0.97	−1.07	−1.34
Fastener	6.7	1.00	−1.10	−1.40
Other**	≤ 10	1.00	−1.10	−1.40
Other**	≥ 500	0.70	−0.80	−0.80

Note: Values are from Figure 6–11A of the Standard. (GC_p) values are obtained using equations in Chapter 2 of this guide. Other C&C can be doors, windows, etc.

Typical calculations of design pressures for girt in Zone 4 are shown below and wall C&C pressures are summarized in **Table G3–24**.

For maximum negative pressure

$p = 18.0[(-0.87) - (\pm 0.18)]$

$p = -18.9$ psf with positive internal pressure (controls)

$p = -12.4$ psf with negative internal pressure

For maximum positive pressure

$p = 18.0[(0.77) - (\pm 0.18)]$

$p = 10.6$ psf with positive internal pressure

$p = 17.1$ psf with negative internal pressure (controls)

Roof Pressures

Effective wind areas of roof C&C (**Table G3–25**).

Purlin

larger of $A = 25(5) = 125 \text{ ft}^2$
or $A = 25(25/3) = 208 \text{ ft}^2$ (controls)

Panel

larger of $A = 5(2) = 10 \text{ ft}^2$ (controls)
or $A = 5(5/3) = 8.3 \text{ ft}^2$

Fastener

$A = 5(1) = 5 \text{ ft}^2$

Table G3-24 Net Controlling Wall Component Pressures (psf)

	Controlling design pressures (psf)			
	Zone 4		*Zone 5*	
C&C	*Positive*	*Negative*	*Positive*	*Negative*
Girt	17.1	−18.9	17.1	−20.0
Panel	20.7	−22.5	20.7	−27.4
Fastener	21.2	−23.0	21.2	−28.4
$A \leq 10$ ft²	21.2	−23.0	21.2	−28.4
$A \geq 500$ ft²	15.8	−17.6	15.8	−17.6

Table G3-25 Roof Coefficients (GC_p), 7° < θ ≤ 27°

		External (GC_p)			
Component	*A (ft²)*	*Zones 1, 2, and 3*	*Zone 1*	*Zone 2*	*Zone 3*
Purlin	208	0.3	−0.8	−1.2	−2.0
Panel	10	0.5	−0.9	−1.7	−2.6
Fastener	5	0.5	−0.9	−1.7	−2.6
Other	≤ 10	0.5	−0.9	−1.7	−2.6
Other	≥ 100	0.3	−0.8	−1.2	−2.0

Typical calculations of design pressures for purlin in Zone 1 are as follows and roof C&C pressures are summarized in **Table G3–26**. For maximum negative pressure.

$p = 18.0[(-0.8) - (\pm 0.18)]$

$p = -17.6$ psf with positive internal pressure (controls)

$p = -11.2$ psf with negative internal pressure

For maximum positive pressure

$p = 18.0[(0.3) - (\pm 0.18)]$

$p = 2.1$ psf with positive internal pressure

$p = 8.6$ psf with negative internal pressure

$p = 10$ psf minimum net pressure (controls) (Section 6.1.4.2 of the Standard)

Table G3–26 Net Controlling Roof Component Pressures (psf)

	Controlling design pressures (psf)			
	Positive	Negative		
Component	Zones 1, 2, and 3	Zone 1	Zone 2	Zone 3
Purlin	10.0*	−17.6	−24.8	−39.2
Panel	12.2	−19.4	−33.8	−50.0
Fastener	12.2	−19.4	−33.8	−50.0
$A \leq 10$ ft²	12.2	−19.4	−33.8	−50.0
$A \geq 500$ ft²	10.0*	−17.6	−24.8	−39.2

* Minimum net pressure controls (Section 6.1.4.2 of the Standard).

Special case of girt that transverses Zones 4 and 5

Width of Zone 5

  smaller of $a = 0.1(200) = 20$ ft
  or $a = 0.4(36.7) = 14.7$ ft (controls)
  but not less than $0.04(200) = 8$ ft
  or   3 ft

Weighted average design pressure

$$P = \frac{14.7(-20.0) + 10.3(-18.9)}{25} = -19.6 \text{ psf}$$

This procedure of using a weighted average may be used for other components and cladding.

Special Case of Strut Purlin (interior)

Strut purlins in the end bay experience combined uplift pressure as a roof component (C&C) and axial load as part of the MWFRS.

Component Pressure

  End bay purlin located in Zones 1 and 2
  Width of Zone 2, $a = 14.7$ ft
  Weighted average design pressure

$$= \frac{14.7(-24.8) + 10.3(-17.6)}{25} = -21.9 \text{ psf}$$

(Purlin in Zones 2 and 3 will have higher pressure)

MWFRS Load

Figure G3–16 shows design pressure on end wall with wind parallel to ridge with positive internal pressure (consistent with high uplift on the purlin). Assuming that the end wall is supported at the bottom and at the roof line, the effective axial load on an end bay purlin can be determined.

Combined Design Load on Interior Strut Purlin

Figure G3–18(a) shows combined design loads on an interior strut purlin. Note that many metal building manufacturers support the top of the wall panels with the eave strut purlin (see **Figure G3–18(b)**). For this case, the eave purlin also serves as a girt, and the negative wall pressures of Zones 5 and 4 would occur for the same wind direction as the maximum negative uplift pressures on the purlin (refer to Zones 3 and 2). Thus, in this instance, the correct load combination would involve biaxial bending loads based on C&C pressures combined with the MWFRS axial load.

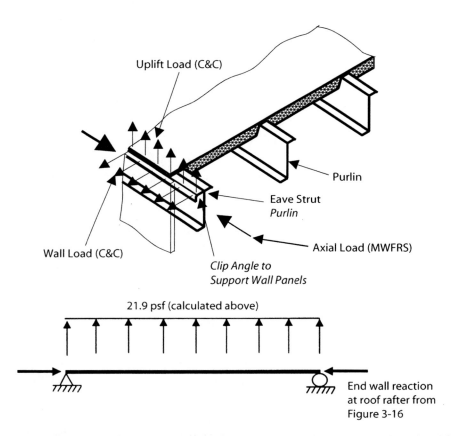

Figure G3–18 Combined Design Loads on Interior Strut Purlin. Note: a) (bottom) combined uplift and axial design loads and b) (top) eave strut purlin supporting roof and wall panels.

3.7 Example 7: Building from Ex. 6 Using Low-Rise Building Provisions

This example illustrates the use of the low-rise building provisions to determine design pressures for the MWFRS. For this purpose, the building used has the same dimensions as the building in Ex. 6 (Section 3.6 of this guide). The design pressures on C&C will be the same as those for Ex. 6. The building is shown in **Figure G3–13**. The building data are as shown here.

Location   Memphis, Tennessee

Terrain   Flat farmland

Dimensions   200 ft × 250 ft in plan
Eave height of 20 ft
Roof slope 4:12 (18.4°)

Framing   Rigid frame spans the 200-ft direction
Rigid frame bay spacing is 25 ft
Lateral bracing in the 250-ft direction is provided by a "wind truss" spanning the 200 ft to side walls and cable/rod bracing in the planes of the walls
Openings uniformly distributed

3.7.1 Low-Rise Building

Section 6.2 of the Standard specifies two requirements for a building to qualify as a low-rise building: (1) mean roof height has to be less than or equal to 60 ft, and (2) mean roof height does not exceed least horizontal dimension. A building with these dimensions qualifies as a low-rise building and the alternate provisions of Section 6.5.12.2.2 may be used.

3.7.2 Exposure, Building Classification, and Basic Wind Speed

Exposure C, same as Ex. 6

Category II, same as Ex. 6

Enclosed building (openings uniformly distributed)

$V = 90$ mph, same as Ex. 6

3.7.3 Velocity Pressure

The low-rise building provisions for MWFRS in the Standard use the velocity pressure at mean roof height, h, for calculation of all external and internal pressures, including the windward wall. All pressures for a given zone are assumed to be uniformly distributed with respect to height above ground.

Mean roof height $h = 36.7$ ft

The velocity pressures are computed using

$$q_h = 0.00256 K_h K_{zt} K_d V^2 I \text{ (psf)} \quad \text{(Eq. 6-15)}$$

where

q_h = Velocity pressure at mean roof height, h
K_h = 1.02 for Exposure C (see Table 6-3 of the Standard)
K_{zt} = 1.0 topographic factor (see Section 6.5.7.1)
K_d = 0.85 (see Table 6-4)
V = 90 mph basic wind speed (see Figure 6-1)
I = 1.0 for Category II (50-yr mean return interval)

therefore,

$$q_h = 0.00256(1.02)(1.0)(0.85)(90)^2(1.0) = 18.0 \text{ psf}$$

3.7.4 Design Pressures for the MWFRS

The equation for the determination of design wind pressures for MWFRS for low-rise buildings is given in Section 6.5.12.2.2 of the Standard.

$$p = q_h[(GC_{pf}) - (GC_{pi})] \quad \text{(Eq. 6-18)}$$

where

q_h = The velocity pressure at mean roof height associated with Exposure C
(GC_{pf}) = The external pressure coefficients from Figure 6-10 of the Standard
(GC_{pi}) = The internal pressure coefficient from Figure 6-5 of the Standard

The building must be designed for all wind directions using the eight loading patterns shown in Figure 6-10 of the Standard. For each of these patterns, both positive and negative internal pressures must be considered, resulting in a total of 16 separate loading conditions. However, if the building is symmetrical, the number of separate loading conditions will be reduced to eight (two directions of MWFRS being designed for normal load and torsional load cases—a total of four load cases, one windward corner, and two internal pressures). The load patterns are applied to each building corner in turn as the reference corner.

External Pressure Coefficients

The roof and wall external pressure coefficients (GC_{pf}) are functions of the roof slope, θ (see **Tables G3–27 and G3–28**).

Table G3–27 Transverse Direction ($\theta = 18.4°$), GC_{pf} by Building Surface

1	2	3	4	5	6	1E	2E	3E	4E
0.52	−0.69	−0.47	−0.42	−0.45	−0.45	0.78	−1.07	−0.67	−0.62

GC_{pf} calculated by linear interpolation.

Table G3–28 Longitudinal Direction (θ = 0°), GC_{pf}, by Building Surface

1	2	3	4	5	6	1E	2E	3E	4E
0.40	−0.69	−0.37	−0.29	−0.45	−0.45	0.61	−1.07	−0.53	−0.43

Width of end zone surface

    smaller of $2a = 2(0.1)(200) = 40$ ft
    or $2(0.4)(36.7) = 29.4$ ft (controls)
    but not less than $2(0.04)(200) = 16$ ft
    or $2(3) = 6$ ft

Internal Pressure Coefficients

Openings are assumed to be evenly distributed in the walls, and since Memphis, Tennessee, is not located in a hurricane-prone region, the building qualifies as an enclosed building (see Section 6.2 of the Standard). The internal pressure coefficients are given from Figure 6-5 as $(GC_{pi}) = \pm 0.18$.

3.7.5 Design Wind Pressures

Design wind pressures in the transverse and longitudinal directions are shown in **Tables G3–29** and **G3–30**.

Calculation for Surface 1

$$p = 18.0\,[0.52 - (\pm 0.18)] = +6.1 \text{ or } +12.6$$

Table G3–29 Design Wind Pressures, Transverse Direction

Building surface	(GC_{pf})	Design pressure (psf)	
		$(+GC_{pi})$	$(-GC_{pi})$
1	0.52	6.1	12.6
2	−0.69	−15.6	−9.2
3	−0.47	−11.7	−5.2
4	−0.42	−10.8	−4.3
5	−0.45	−11.3	−4.9
6	−0.45	−11.3	−4.9
1E	−0.78	10.8	17.3
2E	−1.07	−22.5	−16.0
3E	−0.67	−15.3	−8.8
4E	−0.62	−14.4	−7.9

Table G3-30 Design Wind Pressures, Longitudinal Direction

Building surface	(GC_{pf})	Design pressure (psf) $(+GC_{pi})$	Design pressure (psf) $(-GC_{pi})$
1	0.40	4.0	10.5
2	−0.69	−15.6	−9.2
3	−0.37	−9.9	−3.4
4	−0.29	−8.5	−2.0
5	−0.45	−11.3	−4.9
6	−0.45	−11.3	−4.9
1E	0.61	7.7	14.2
2E	−1.07	−22.5	−16.0
3E	−0.53	−12.8	−6.3
4E	−0.43	−11.0	−4.5

Application of Pressures on Building Surfaces 2 and 3

Note 8 of Figure 6-10 of the Standard states that when the roof pressure coefficient, GC_{pf}, is negative in Zone 2, it shall be applied in Zone 2 for a distance from the edge of the roof equal to 0.5 times the horizontal dimension of the building measured parallel to the direction of the MWFRS being designed or $2.5h$, whichever is less. The remainder of Zone 2 that extends to the ridge line shall use the pressure coefficient GC_{pf} for Zone 3. Thus, the distance from the edge of the roof is the smaller of

$$0.5(200) = 100 \text{ ft for transverse direction}$$
$$0.5(250) = 125 \text{ ft for longitudinal direction}$$
or $$(2.5)(36.7) = 92 \text{ ft for both directions (controls)}$$

Therefore, Zone 3 applies over a distance of 105 − 92 = 13 ft in what is normally considered to be Zone 2 (adjacent to ridge line) for transverse direction and 125 − 92 = 33 ft for longitudinal direction.

3.7.6 Loading Cases

Because the building is symmetrical, the four loading cases provide all the required combinations provided the design is accomplished by applying loads for each of the four corners. The load combinations illustrated in **Figures G3–19** through **G3–22** are to be used to design the rigid frames, the "wind truss" spanning across the building in the 200-ft direction, and the rod/cable bracing in the planes of the walls (see **Figure G3–13** in Section 3.6 of this guide).

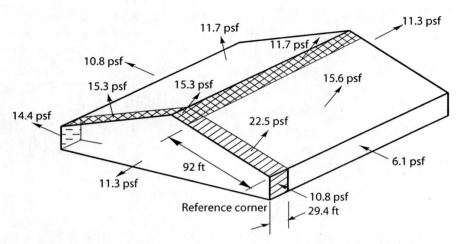

Figure G3–19 Design Pressures for Transverse Direction with Positive Internal Pressure. Note: The pressures are assumed to be uniformly distributed over each of the surfaces shown.

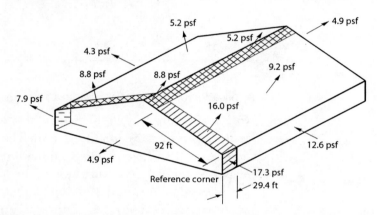

Figure G3–20 Design Pressures for Transverse Direction with Negative Internal Pressure. Note: The pressures are assumed to be uniformly distributed over each of the surfaces shown.

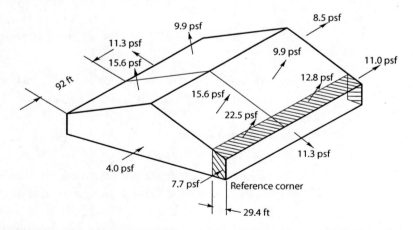

Figure G3–21 Design Pressures for Longitudinal Direction with Positive Internal Pressure. Note: The pressures are assumed to be uniformly distributed over each of the surfaces shown.

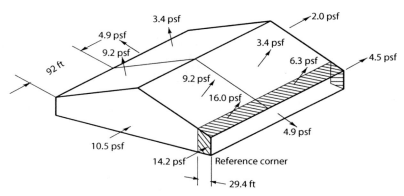

Figure G3–22 Design Pressures for Longitudinal Direction with Negative Internal Pressure. Note: The pressures are assumed to be uniformly distributed over each of the surfaces shown.

Torsional Load Cases

Since the mean roof height, $h = 36.7$ ft, is greater than 30 ft and if the roof diaphragm is assumed to be rigid, torsional load cases need to be considered (see exception in Note 5 in Figure 6-10 of the Standard if building is designed with flexible diaphragm). Pressures in "T" zones are 25% of the full design pressures; the "T" zones are shown in Figure 6-10 of the Standard. Other surfaces will have the full design pressures. The "T" zone pressures with positive and negative internal pressures for transverse and longitudinal directions are shown in **Tables G3–31** and **G3–32**, respectively.

Figures G3–19 through **G3–26** show design pressure cases for one reference corner; these cases are to be considered for each corner.

3.7.7 Design Wind Pressures for C&C

The design pressures for C&C are the same as shown for Ex. 6 (Section 3.6 of this guide).

Table G3–31 Design Wind Pressure for Zone "T," Transverse Direction

Building surface	Design pressures (psf)	
	$(+GC_{pi})$	$(-GC_{pi})$
1T	1.5	3.2
2T	−3.9	−2.3
3T	−2.9	−1.3
4T	−2.7	−1.1

Table G3–32   Design Wind Pressure for Zone "T," Longitudinal Direction

Building surface	Design pressures (psf)	
	$(+GC_{pi})$	$(-GC_{pi})$
1T	1.0	2.6
2T	−3.9	−2.3
3T	−2.5	−0.9
4T	−2.1	−0.5

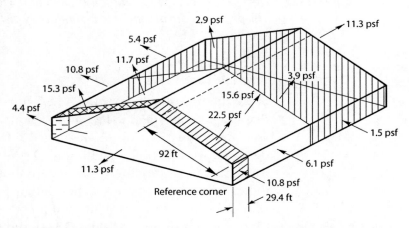

Figure G3–23   *Torsional Load Case for Transverse Direction with Positive Internal Pressure. Notes: The pressures are assumed to be uniformly distributed over each of the surfaces shown. Roof pressures of 22.5, 15.6, and 3.9 psf apply up to 92 ft; the remaining 13 ft up to the ridge line will have pressure of 15.3, 11.7, and 2.9 psf.*

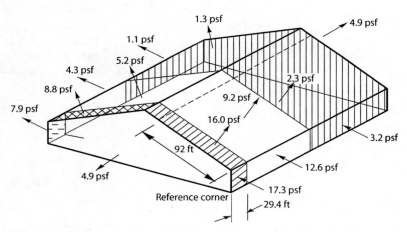

Figure G3–24   *Torsional Load Case for Transverse Direction with Negative Internal Pressure. Notes: The pressures are assumed to be uniformly distributed over each of the surfaces shown. Roof pressures of 16.0 9.2, and 2.3 psf apply up to 92 ft; the remaining 13 ft up to the ridge line will have pressure of 8.8, 5.2, and 1.3 psf.*

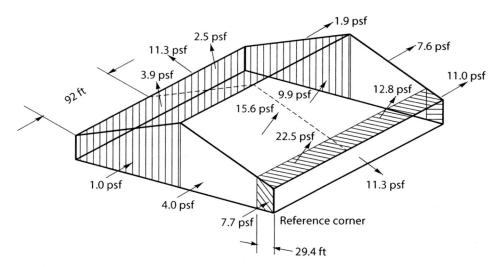

Figure G3–25 Torsional Load Case for Longitudinal Direction with Positive Internal Pressure. Note: The pressures are assumed to be uniformly distributed over each of the surfaces shown.

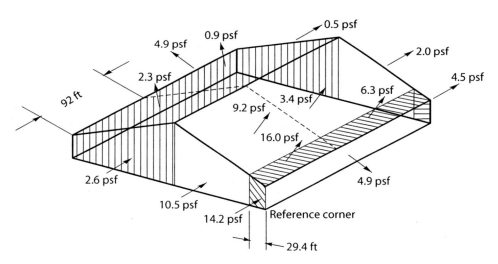

Figure G3–26 Torsional Load Case for Longitudinal Direction with Negative Internal Pressure. Note: The pressures are assumed to be uniformly distributed over each of the surfaces shown.

3.8 Example 8: 40-ft × 80-ft Commercial Building with Monoslope Roof with Overhang

In this example, design pressures for a typical retail store in a strip-mall are determined. The building's dimensions are shown in **Figure G3–27**. The building data are as shown here.

Location Boston, Massachusetts, within 1 mi of the coastal mean high watermark

Topography Homogeneous

Terrain Suburban

Dimensions 40 ft × 80 ft in plan
Monoslope roof with slope of 14° and overhang of 7 ft in plan
Wall heights are 15 ft in front and 25 ft in rear

Framing Walls of CMU on all sides supported at top and bottom; steel framing in front (80-ft width) to support window glass and doors. Roof joists span 41.2 ft with 7.2-ft overhang spaced at 5 ft on center

Cladding Glass and door sizes vary; glazing is debris-impact–resistant and occupies 50% of front wall (80 ft in width)
Roof panels are 2 ft wide and 20 ft long

3.8.1 Building Classification, Enclosure Classification, and Exposure Category

The building is not an essential facility, nor is it likely to be occupied by more than 300 persons at any one time. Use Category II (see Table 1-1 of the Standard). Importance Factor $I = 1.00$ (see Table 6-1 of the Standard).

The building is sited in a suburban area and satisfies the criteria for Exposure B (see Section 6.5.6 of the Standard).

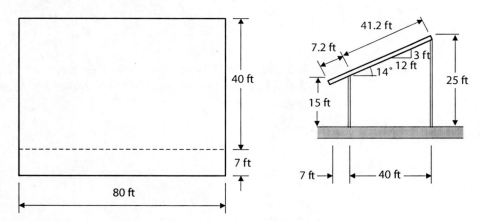

Figure G3–27 *Building Characteristics for Example 8, Commercial Building with Monoslope Roof and Overhang*

The building is sited in a wind-borne debris region. It has glazing (that must be impact resistant) occupying 50% of a wall that receives positive pressure. The building should be classified as enclosed (see Sections 6.5.9.3 and 6.2 of the Standard). The Standard does not require that the building be classified as enclosed if it is located in a wind-borne debris region, just that the openings are required to be protected with impact-resistant glazing or are protected. The wind-borne debris region is defined in Section 6.2.

The building does not meet the requirements of Method 1, Simplified Procedure (Section 6.4 of the Standard), because the roof is neither flat nor gabled. Therefore, Method 2, Analytical Procedure, is used (see Section 6.5.3 of the Standard). The roof is not gabled; hence, the low-rise building provisions may not be used.

The values in Figure 6-10 were obtained from wind tunnel studies of rigid, gable-framed buildings. Their use for a monoslope roof requires considerable judgment. The design examples presented in Ex. 7 (Section 3.7 of this guide) illustrate use of the pressure coefficients of Figure 6-10, and the Commentary in the Standard gives the background for (GC_{pf}) values.

3.8.2 Basic Wind Speed

The wind speed contour of 110 mph traverses over Boston, Massachusetts (Figure 6-1c of the Standard); use a basic wind speed of 110 mph.

3.8.3 Velocity Pressures

The velocity pressures (**Table G3–33**) are calculated using the following equation (see Section 6.5.10 of the Standard):

$$q = 0.00256 K_z K_{zt} K_d V^2 I \text{ (psf)}$$
$$= 0.00256 K_z (1.0)(0.85)(110)^2(1.0)$$
$$= 26.33 K_z \text{ (psf) (Eq. 6-15)}$$

Table G3–33 Velocity Pressures, q_z, q_i, and q_h (psf)

Height (ft)	MWFRS		C&C	
	Exposure B, Case 2	q_z, q_i	Exposure B, Case 1	q_h
0–15	0.57	15.01		
h = 20	0.62	16.32*	0.70	18.43
25	0.66	17.38		

* q_h = 16.32 psf for MWFRS.

where

K_z = Value obtained from Table 6-3 of the Standard

K_{zt} = 1.0 homogeneous terrain

I = 1.0 for Category II building (see Table 6-1)

K_d = 0.85, see Table 6-4

The provisions of the Standard require the use of the external pressure coefficients, C_p, from Figure 6-6; hence, the exposure coefficients, K_z, are based on Exposure B, Case 2, for MWFRS and Exposure B, Case 1, for C&C (see Table 6-3).

3.8.4 Design Pressures for MWFRS

The equation for rigid buildings of all heights is given in Section 6.5.12.2 of the Standard.

$$p = qGC_p - q_i(GC_{pi}) \qquad \text{(Eq. 6-17)}$$

where

$q = q_z$ for windward wall

$q_i = q_h$ for windward and leeward walls, side walls and roof

G = Value determined from Section 6.5.8 of the Standard

C_p = Value obtained from Figure 6-6 of the Standard

(GC_{pi}) = Value obtained from Figure 6-5 of the Standard

For positive internal pressure evaluation, the Standard permits q_i to be conservatively evaluated at height h ($q_i = q_h$).

Gust Effect Factor

The gust effect factor for non-flexible (rigid) buildings is given in Section 6.5.8 of the Standard as $G = 0.85$. The size of the building would not permit a reduction in G based on Eq. 6-4 of the Standard.

Wall External Pressure Coefficients

The coefficients for the windward and side walls in **Table G3–34** are given in Figure 6-6 of the Standard as $C_p = +0.8$ and -0.7, respectively. The values for the leeward wall depend on L/B; they are different for the two directions: (1) wind parallel to roof slope (normal to ridge), and (2) wind normal to roof slope (parallel to ridge).

Roof External Pressure Coefficients

Since the building has a monoslope roof, the roof surface for wind directed parallel to the slope (normal to ridge) may be a windward or a leeward surface.

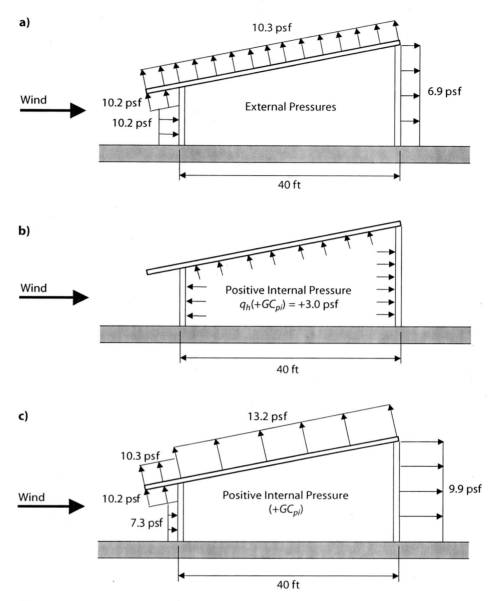

Figure G3–28   Design Pressures for MWFRS for Wind Parallel to Roof Slope, Normal to 15-ft Wall, and Positive Internal Pressure. Note: a) external pressures, b) positive internal pressure, and c) combined external and positive internal pressure.

Effective wind area

For span of 15 ft, $A = 15(15/3) = 75 \text{ ft}^2$

For span of 20 ft, $A = 20(20/3) = 133 \text{ ft}^2$

For span of 25 ft, $A = 25(25/3) = 208 \text{ ft}^2$

Table G3–36 Design Pressures for MWFRS: Wind Parallel to Roof Slope, normal to ridge line

Wind direction	Surface	Z (ft)	q_z (psf)	Gust effect	External C_p*	Design pressure (psf) $(+GC_{pi})$	Design pressure (psf) $(-GC_{pi})$
Windward wall (15 ft)	Windward wall	0–15	15.01	0.85	0.80	7.3	13.2
	Leeward wall	0–25	16.32	0.85	–0.50	–9.9	–3.9
	Side wall	All	16.32	0.85	–0.70	–12.6	–6.8
	Roof	—	16.32	0.85	–0.74	–13.2	–7.3
	Overhang top	—	16.32	0.85	–0.74	–10.3†	–10.3†
	Overhang bottom	—	15.01	0.85	0.80	10.2†	10.2†
Windward wall (25 ft)	Windward wall	0–15	15.01	0.85	0.80	7.3	13.2
		15–20	16.32	0.85	0.80	8.2	14.0
		20–25	17.38	0.85	0.80	8.9	14.8
	Leeward wall	All	16.32	0.85	–0.50	–9.9	–3.9
	Side wall	All	16.32	0.85	–0.70	–12.6	–6.8
	Roof	—	16.32	0.85	–0.50	–9.9	–4.0
	Overhang top	—	16.32	0.85	–0.50	–6.9†	–6.9†
	Overhang bottom	—	—	—	—	0.0†	0.0†

* External pressure calculations include $G = 0.85$.
† Overhang pressures are not affected by internal pressures. The Standard does not address bottom surface pressures for leeward overhang. It could be argued that leeward wall pressure coefficients can be applied, but note that neglecting the bottom overhang pressures would be conservative in this application.

(GC_{pi}) = +0.18 and –0.18, previously determined from Figure 6-5 of the Standard

Wall Design Pressures

Wall external pressure coefficients are presented in **Table G3–38**. Since the CMU walls are supported at the top and bottom, the effective wind area will depend on the span length.

Table G3–34   Wall Pressure Coefficients (C_p)

Surface	Wind direction	L/B	C_p
Leeward wall	∥ to roof slope	0.5	−0.5
Leeward wall	⊥ to roof slope	2.0	−0.3
Windward wall	—	—	0.8
Side walls	—	—	−0.7

The value of $h/L = 0.5$ in this case, and the proper coefficients are obtained from linear interpolation for θ = 14° (see **Table G3–35**).

When wind is normal to the roof slope (parallel to ridge), angle θ = 0 and $h/L = 0.25$.

For the overhang, Section 6.5.11.4.1 of the Standard requires $C_p = 0.8$ for wind directed normal to 15-ft wall. The Standard does not address the leeward overhang for the case of wind directed toward 25-ft wall and perpendicular to roof slope (parallel to ridge). A $C_p = -0.5$ could be used (coefficient for leeward wall), but the coefficient has been conservatively taken as 0.

The building is sited in a hurricane-prone region less than 1 mi from the coastal mean high-water level. The basic wind speed is 110 mph and the glazing must be designed to resist wind-borne debris impact (or some other method of protecting the glazing is installed, such as shutters). Thus, as noted earlier, the building is classified as enclosed, for this example. The internal pressure coefficients, from Figure 6-5 of the Standard, are as follows (GC_{pi}) = ± 0.18.

Typical Calculations of Design Pressures for MWFRS

For cases with wind parallel to slope with 15-ft windward wall (**Table G3–36**).

Pressure on Leeward Wall

$$p = q_h GC_p - q_h(\pm GC_{pi})$$
$$= 16.3(0.85)(-0.5) - (16.3)(+0.18) = -9.9 \text{ psf with positive internal pressure}$$
$$\text{and} = 16.3(0.85)(-0.5) - (16.3)(-0.18) = -4.0 \text{ psf with negative internal pressure}$$

Pressure on Overhang Top Surface

$$p = q_h GC_p = 16.3(0.85)(-0.74) = -10.3 \text{ psf}$$

Pressure on Overhang Bottom Surface (same as windward wall external pressure)

$$p = q_z GC_p = 15.0(0.85)(0.8) = 10.2 \text{ psf}$$

Note that q_z was evaluated for $z = 15$ ft for bottom surface of overhang as C_p coefficient is based on induced pressures at top of wall.

Table G3–35   Roof Pressure Coefficients (C_p)

Wind direction	h/L	θ°	C_p
∥ to roof slope	0.5	14	−0.74, −0.18* as windward slope
∥ to roof slope	0.5	14	−0.50 as leeward slope
⊥ to roof slope	0.25	0	
distance from windward edge			
0–80 ft*			−0.18†
0–20 ft*			−0.90
20–40 ft			−0.50
40–80 ft			−0.30

* Distance from the windward edge of the roof.
† The values of smaller uplift pressures on the roof can become critical when wind load is combined with roof live load or snow load; load combinations are given in Sections 2.3 and 2.4 of the Standard. For brevity, loading for this value is not shown here.

Figures G3–28 and **G3–29** illustrate the external, internal, and combined pressure for wind directed normal to the 15-ft wall. **Figures G3–30** and **G3–31** illustrate combined pressure for wind directed normal to the 25-ft wall and perpendicular to slope (parallel to ridge line), respectively (**Table G3–37**).

3.8.5 Design Wind Load Cases

Section 6.5.12.3 of the Standard requires that any building whose wind loads have been determined under the provisions of Sections 6.5.12.2.1 and 6.5.12.2.3 shall be designed for wind load cases as defined in Figure 6-9 of the Standard. Case 1 includes the loadings shown in **Figures G3–28** through **G3–31**. The exception in Section 6.5.12.3 of the Standard indicates that a combination of windward (P_w) and leeward (P_L) loads is applied for Load Cases 3 only since mean roof height h of the building is less than 30 ft.

3.8.6 Design Pressures for C&C

The design pressure equation for C&C for building with mean roof height $h \leq 60$ ft is given in Section 6.5.12.4.1 of the Standard.

$$P = q_h[(GC_p) - (GC_{pi})] \quad \text{(Eq. 6-22)}$$

where

q_h = Velocity pressure at mean roof height associated with Exposure B, Case 1 (q_h = 18.43 psf, previously determined)

(GC_p) = External pressure coefficients from Figures 6-11A, 6-11C, and 6-14B of the Standard

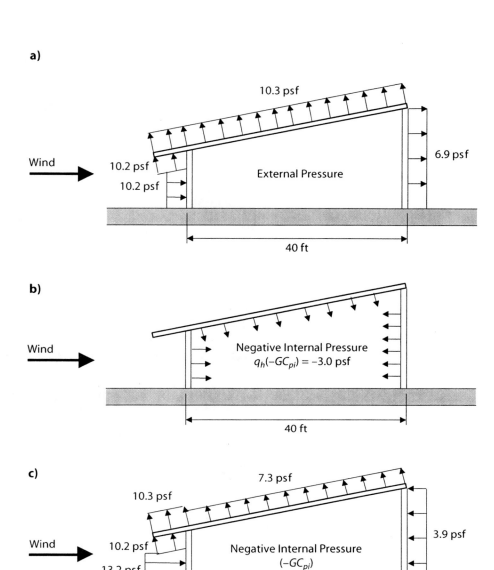

Figure G3-29 Design Pressures for MWFRS for Wind Parallel to Roof Slope, Normal to 15-ft Wall, and Negative Internal Pressure. Note: a) external pressures, b) negative internal pressure, and c) combined external and negative internal pressure.

Width of Zone 5 (Figure 6-11A)

    smaller of $a = 0.1(40) = 4$ ft (controls)
    or $a = 0.4(20) = 8$ ft
    but not less than $a = 0.4(40) = 1.6$ ft
    or $a = 3$ ft

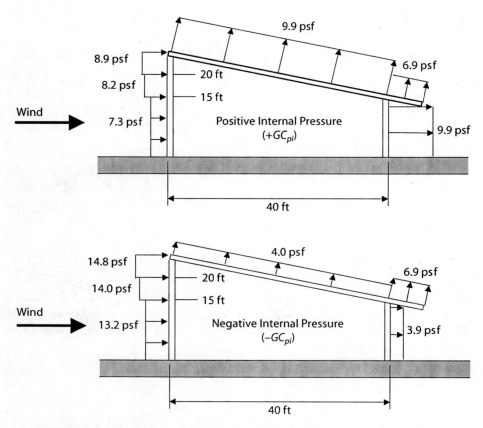

Figure G3–30 Combined Design Pressures for MWFRS for Wind Parallel to Roof Slope, Normal to 25-ft Wall

Design pressures are the critical combinations when the algebraic sum of the external and internal pressures is a maximum.

Typical Calculations for Design Pressures for 15-ft Wall, Zone 4

Wall design pressures are presented in **Table G3–39**.

$$p = q_h[(GC_p) - (\pm GC_{pi})]$$
$$= 18.43[(0.85) - (-0.18)] = 19.0 \text{ psf}$$
$$\text{and} = 18.43[(-0.95) - (0.18)] = -20.8 \text{ psf}$$

The CMU walls are designed for pressures determined for Zones 4 and 5 using appropriate tributary areas.

The design pressures for doors and glazing can be assessed by using appropriate pressure coefficients associated with their effective wind areas.

Roof Design Pressures

Roof external pressure coefficients are presented in **Table G3–40**; roof design pressures are presented in **Table G3–41**.

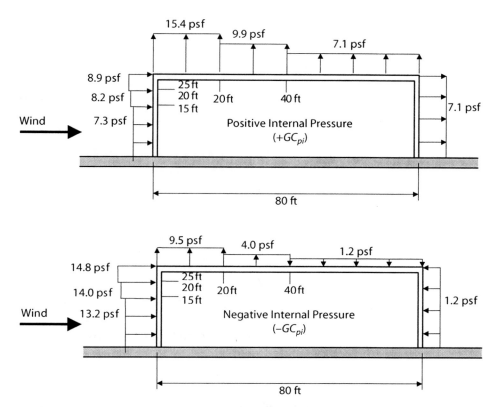

Figure G3–31 Combined Design Pressures for MWFRS for Wind Perpendicular to Roof Slope, Parallel to Ridge Line

Table G3–37 Design Pressures for MWFRS: Wind Normal to Roof Slope, parallel to ridge line

Surface	z or distance (ft)*	q_z† (psf)	Gust effect, G	C_p	Design pressure (psf) +(GC_{pi})‡	Design pressure (psf) −(GC_{pi})†
Windward wall	0–15	15.01	0.85	0.8	7.3	13.2
	15–20	16.32	0.85	0.8	8.2	14.0
	20–25	17.38	0.85	0.8	8.9	14.8
Leeward wall	All	16.32	0.85	−0.3	−7.1	1.2
Side wall	All	16.32	0.85	−0.7	−12.6	−6.8
Roof**	0–20	16.32	0.85	−0.9	−15.4	−9.5
	20–40	16.32	0.85	−0.5	−9.9	−4.0
	40–80	16.32	0.85	−0.3	−7.1	1.2

*   External pressure calculations include $G = 0.85$.
†   Internal pressure is associated with $q_h = 16.32$ psf.
‡   Distance along roof is from leading windward edge.
** Pressure on overhang is only external pressure (contribution on underside is conservatively neglected).

Wind Loads: Guide to the Wind Load Provisions of ASCE 7-05

Table G3-38 Wall External Pressure Coefficients (GC_p) by Zone

A (ft²)	Zones 4 and 5 (+GC_p)	Zone 4 (−GC_p)	Zone 5 (−GC_p)
75	0.85	−0.95	−1.09
133	0.80	−0.90	−1.00
208	0.77	−0.87	−0.93

Table G3-39 Wall Design Pressures (psf) by Zone

Wall height (ft)	Zones 4 and 5 Positive	Zone 4 Negative	Zone 5 Negative
15	19.0	−20.8	−23.4
20	18.1	−19.9	−21.7
25	17.5	−19.4	−20.4

Note: q_h = 18.43 psf.

Table G3-40 Roof External Pressure Coefficients (GC_p), θ = 14°, by Zone

Component	A (ft²)	Zones 1, 2, and 3 (+GC_p)	Zone 1 (−GC_p)	Zone 2 (−GC_p)	Zone 3 (−GC_p)
From Figure 6–14B of the Standard					
Joist panel	566	0.3	−1.1	−1.2	−2.0
	10	0.4	−1.3	−1.6	−2.9
From Figure 6–11C of the Standard					
Joist panel	566	0.3	−0.8	−2.2*	−2.5*
	10	0.5	−0.9	−2.2*	−3.7*

* Values are from overhang chart in Figure 6–11C of the Standard

Table G3-41 Roof Design Pressures (psf) by Zone

Component	Zones 1, 2, and 3* Positive	Zone 1 Negative	Zone 2 Negative	Zone 3 Negative
Joist	10.0	−23.6	−25.4	−40.2
Joist overhang	10.0†	−14.7	−40.6	−46.1
Panel	10.7	−27.3	−32.8	−56.8
Panel in overhang	10.0†	−19.9	−43.9	−71.5

Note: q_h = 18.43 psf.

* Zones for overhang are in accordance with Figure 6–11C of the Standard.
† Section 6.1.4.2 of the Standard requires a minimum of 10 psf.

Effective wind area

Roof joist

$$A = (41.2)(5) = 206 \text{ ft}^2$$

$$\text{or } A = (41.2)(41.2/3) = 566 \text{ ft}^2 \text{ (controls)}$$

Roof panel

$$A = (5)(2) = 10 \text{ ft}^2 \text{ (controls)}$$

$$\text{or } A = (5)(5/3) = 8.3 \text{ ft}^2$$

Had the effective wind area of the roof joist been greater than 700 ft^2, its external pressure coefficients (GC_p) would still have been determined on the basis of components and cladding. The statement in Section 6.5.12.1.3 of the Standard, in which provisions for MWFRS may be used for a major component, is valid only when the tributary area is greater than 700 ft^2. The tributary area for the roof joist is 242 ft^2.

Section 6.5.11.4.2 of the Standard requires that pressure coefficients for components and cladding of roof overhangs be obtained from Figure 6-11C. Note that the zones for roof overhangs in Figure 6-11C are different from the zones for a monoslope roof in Figure 6-14B.

Width of zone distance a

smaller of $\quad a = 0.1(40) = 4 \text{ ft (controls)}$
or $\quad a = 0.4(20) = 8 \text{ ft}$
but not less than $\quad a = 0.4(40) = 1.6 \text{ ft}$
or $\quad a = 3 \text{ ft}$

The widths and lengths of Zones 2 and 3 for a monoslope roof are shown in Figure 6-14B of the Standard (they vary from a to 4a); for overhangs, widths and lengths are shown in Figure 6-11C.

Similar to the determination of design pressures for walls, the critical design pressures for roofs are the algebraic sum of the external and internal pressures. The design pressures for overhang areas are based on pressure coefficients obtained from Figure 6-11C of the Standard.

Typical Calculations for Joist Pressures

Zone 2

$$p = q_h[(GC_p) - (\pm GC_{pi})]$$
$$= 18.43[(0.3) - (-0.18)] = 8.8 \text{ psf}$$
$$\text{and } = 18.43[(-1.2) - (0.18)] = -25.4 \text{ psf}$$

Zones for the monoslope roof and for overhang are shown in **Figure G3–32**. The panels are designed for the pressures indicated.

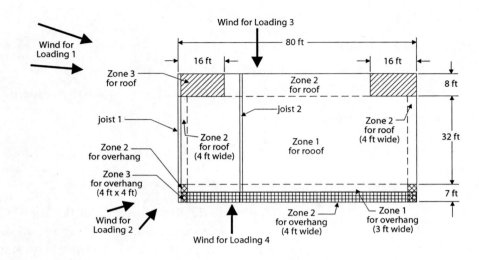

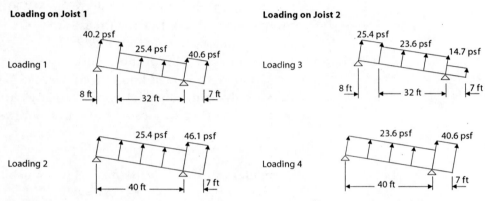

Figure G3–32 *Design Pressures for Typical Joists and Pressure Zones for Roof Components and Cladding*

Roof joist design pressures need careful interpretation. The high pressures in corner or eave areas do not occur simultaneously at both ends. Two loading cases: wind loadings 1, 2 for joist 1 and wind loadings 3, 4 for joist 2, are shown in **Figure G3–32** based on the following zones:

- Joist 1, loading 1: Zones 2 and 3 for roof and Zone 2 for overhang
- Joist 1, loading 2: Zone 2 for roof and Zones 2 and 3 for overhang
- Joist 2, loading 3: Zones 1 and 2 for roof and Zone 1 for overhang
- Joist 2, loading 4: Zone 1 for roof and Zones 1 and 2 for overhang

For simplicity, only one zone is used for overhang pressures in **Figure G3–32**.

3.9 Example 9: U-Shaped Apartment Building

This example demonstrates calculation of wind loads for a U-shaped apartment building, shown in **Figure G3–33**. Data for the building are as shown here.

Location Birmingham, Alabama

Topography Homogeneous

Terrain Suburban

Dimensions 170 ft × 240 ft overall in plan
Roof eave height of 30 ft
Hip roof with 5 on 12 pitch

Framing Typical timber construction
Wall studs are spaced at 16 in. on center, 10 ft tall
Roof trusses are spaced at 24 in. on center, spanning 70 ft between interior or exterior bearing walls
Floor slab and roof sheathing provide diaphragm action

Cladding Location is outside a wind-borne debris region, so no glazing protection is required. Window units are 3 ft × 4 ft

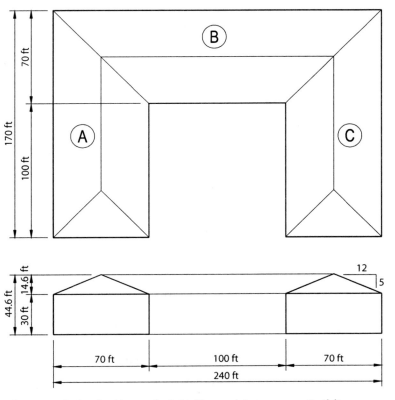

Figure G3–33 *Building Characteristics for Example 9, U-Shaped Apartment Building*

Wind Loads: Guide to the Wind Load Provisions of ASCE 7-05

The building is non-symmetrical, and therefore is ineligible for design by Method 1, Simplified Procedure, of ASCE 7-05. Method 2, Analytical Procedure is used. The building is less than 60 ft tall, so it is possible to use low-rise provisions of Section 6.5.12.2.2. However, because U-, T-, and L-shaped buildings are not specifically covered, the adaptation of the low-rise "pseudo pressure" coefficients to buildings outside the scope of the research is not recommended. Therefore, use the "all heights method" of Section 6.5.12.2.1 of the Standard.

3.9.1 Exposure

The building is located in a suburban area; according to Section 6.5.6.3 of the Standard, Exposure B is used.

3.9.2 Building Classification

The building function is residential. It is not considered an essential facility, nor is it likely to be occupied by 300 persons in a single area at one time. Therefore, building Category II is appropriate (see Table 1-1 of the Standard).

3.9.3 Enclosure

The building is designed to be enclosed. It is not located within a windborne debris region, so glazing protection is not required.

3.9.4 Basic Wind Speed

Selection of the basic wind speed is addressed in Section 6.5.4 of the Standard. Birmingham, Alabama, is located just inside the 90-mph contour; therefore, the basic wind speed $V = 90$ mph (see Figure 6-1b of the Standard).

3.9.5 Velocity Pressures

The velocity pressures are computed using

$$q_z = 0.00256 K_z K_{zt} K_d V^2 I \text{ psf} \qquad \text{(Eq. 6-15)}$$

where

K_z = Value obtained from Table 6-3 of the Standard: Case 1 for C&C and Case 2 for MWFRS

K_{zt} = 1.0 for homogeneous topography

K_d = 0.85 for buildings (see Table 6-4 of the Standard)

V = 90 mph

I = 1.0 for Category II classification (see Table 6-1 of the Standard)

therefore,

$$q_z = 0.00256 K_z (1.0)(0.85)(90)^2(1.0) = 17.63 K_z \text{ psf}.$$

Values for K_z and the resulting velocity pressures are given in **Table G3–42**. The mean roof height is the average of the eave and the peak.

$h = 30 + (14.6/2) = 37.3$ ft

At the mean roof height, $h = 37.3$ ft; the velocity pressure is $q_h = 13.2$ psf.

External Pressure Coefficients

The values for the external pressure coefficients for the various surfaces (**Tables G3–43** through **G3–46**) are obtained from Figure 6-6 of the Standard for each of the surfaces in **Figure G3–34**. The determination of certain pressure coefficients is based on aspect ratios. Even though this U-shaped building will be broken into pieces for the application of pressures, the overall dimensions have greater influence on the MWFRS pressure coefficients than the dimensions of the individual pieces. Therefore, the overall dimensions L and B are used.

When the wind is normal to wall W2, the wind blows over the "A" wing, crosses the courtyard in the middle of the U, and strikes the "C" wing. Although some reduction in the pressures on the "C" wing may occur due to the shielding offered by "A", it is impossible to predict without a wind tunnel study. Therefore, the pressures on the "C" wing are taken as the same as on the "A" wing. There would also be reductions in pressures on the "C" wing if the "A" wing was taller but, again, the amount of the reduction is not possible to predict without a wind tunnel study. If the wind impacts the "A" wing at an angle such that the wind would blow directly into the courtyard, then the "C" wing could still be impacted by the full force of the wind; therefore for this example, the judgment has been made that when the wind is normal to wall W2, wall W6 is also a windward wall (and likewise if the wind is normal to wall W4 where wall W8 is also impacted).

For wind normal to surface W2 or W4

$L/B = 240/170 = 1.41$

$h/L = 37.3/240 = 0.16$

$\theta = 22.6°$ for a 5-in-12 slope

For wind normal to surface W3 or W1-W7-W5

$L/B = 170/240 = 0.71$

$h/L = 37.3/170 = 0.22$

$\theta = 22.6°$ for a 5-in-12 slope

The windward wall C_p is always 0.8, the side walls are -0.7, and the leeward wall varies with the aspect ratio L/B.

The roof C_p for wind normal to a ridge varies with roof angle and aspect ratio, h/L. $h/L \le 0.25$ for all wind directions. The roof angle θ is

Table G3–42 q_z Velocity Pressures

Height (ft)	MWFRS		C&C	
	K_z	q_z (psf)	K_z	q_z (psf)
0–15	0.57	10.1	0.70	12.3
20	0.62	10.9	0.70	12.3
30	0.70	12.3	0.70	12.3
Mean roof ht = 37.3	0.75	13.2	0.75	13.2

Table G3–43   External Pressure Coefficients (C_p) for Wind Normal to Wall W2

Surface type	Surface designation	Surface	Case	L/B or h/L	C_p
Walls	W2, W6	Windward		All	+0.80
	W4, W8	Leeward		1.41	−0.42
	W1, W3, W5, W7	Side		All	−0.70
Roofs (⊥ to ridge)	A1, C2	Windward	Negative	0.16	−0.25
			Positive	0.16	+0.25
	A2, C1	Leeward		0.16	−0.60
Roofs (∥ to ridge)	A3, C3	Side	0 to h	0.16	−0.90*
			h to $2h$	0.16	−0.50*
	B1, B2	Side	0 to h	0.16	−0.90*
			h to $2h$	0.16	−0.50*
			> $2h$	0.16	−0.30*

* The values of smaller uplift pressures ($C_p = -0.18$) on the roof can become critical when wind load is combined with roof live load or snow load; load combinations are given in Sections 2.3 and 2.4 of the Standard. For brevity, loading for this value is not shown here.

always 22.6°, so interpolate between 20° and 25°. The C_p for wind parallel to a ridge varies with h/L and with distance from the leading edge of the roof.

3.9.6 Design Wind Pressures for the MWFRS

The design pressures for this building are obtained by the equation

$$p = qGC_p - q_i(GC_{pi}) \qquad \text{(Eq. 6-17)}$$

where

$q = q_z$ for windward wall at height z above ground

$q = q_h = 13.2$ psf for leeward wall, side walls, and roof

Table G3–44 External Pressure Coefficients (C_p) for Wind Normal to Wall W4

Surface type	Surface designation	Surface	Case	L/B or h/L	C_p
Walls	W4, W8	Windward		All	+0.80
	W6, W2	Leeward		1.41	−0.42
	W1, W3, W5, W7	Side		All	−0.70
Roofs (⊥ to ridge)	C1, A2	Windward	Negative	0.16	−0.25
			Positive	0.16	+0.25
	C2, A1	Leeward		0.16	−0.60
Roofs (∥ to ridge)	A3, C3	Side	0 to h	0.16	−0.90
			h to $2h$	0.16	−0.50
	B1, B2	Side	0 to h	0.16	−0.90
			h to $2h$	0.16	−0.50
			> $2h$	0.16	−0.30

Table G3–45 External Pressure Coefficients (C_p) for Wind Normal to Wall W3

Surface type	Surface designation	Surface	Case	L/B or h/L	C_p
Walls	W3	Windward		All	+0.80
	W1, W7, W5	Leeward		0.71	−0.50
	W2, W4, W6, W8	Side		All	−0.70
Roofs (⊥ to ridge)	B1	Windward	Negative	0.22	−0.25
			Positive	0.22	+0.25
	A3, B2, C3	Leeward		0.22	−0.60
Roofs (∥ to ridge)	A1, A2, C1, C2	Side	0 to h	0.22	−0.90
			h to $2h$	0.22	−0.50
			> $2h$	0.22	−0.30

$q_i = q_h$ = 13.2 psf for all surfaces since the building is enclosed

G = 0.85, the gust effect factor for rigid buildings and structures

C_p = External pressure coefficient for each surface as shown in **Tables G3–43** through **G3–46**

(GC_{pi}) = ±0.18, the internal pressure coefficient for enclosed buildings

For windward walls

$$p = q_z GC_p - q_h(GC_{pi}) = q_z(0.85)C_p - 13.2(\pm 0.18) = 0.85 q_z C_p \pm 2.4$$

Table G3–46 External Pressure Coefficients (C_p) for Wind Normal to Wall W1-W7-W5

Surface type	Surface designation	Surface	Case	L/B or h/L	C_p
Walls	W1, W7, W5	Windward		All	+0.80
	W3	Leeward		0.71	−0.50
	W2, W4, W6, W8	Side		All	−0.70
Roofs (⊥ to ridge)	A3, B2, C3	Windward	Negative	0.22	−0.25
			Positive	0.22	+0.25
	B1	Leeward		0.22	−0.60
Roofs (∥ to ridge)	A1, A2, C1, C2	Side	0 to h	0.22	−0.90
			h to $2h$	0.22	−0.50
			$> 2h$	0.22	−0.30

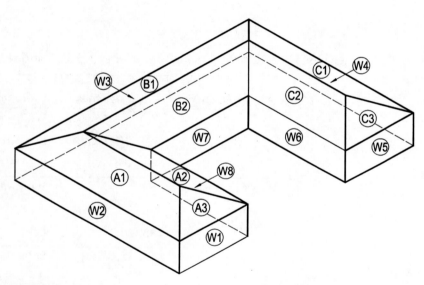

Figure G3–34   Surface Designations

For all other surfaces

$$p = q_h GC_p - q_h(GC_{pi}) = 13.2(0.85)C_p - 13.2(\pm 0.18) = 11.2 C_p \pm 2.4$$

3.9.7 Design Wind Load Cases

Section 6.5.12.3 of the Standard requires that any building whose wind loads have been determined under the provisions of Sections 6.5.12.2.1 and 6.5.12.2.3 shall be designed for wind load cases as defined in Figure 6-9. Case 1 includes the loadings determined in this example and shown in **Tables G3–47** through **G3–50**. A combination of windward (P_W) and leeward (P_L) loads are applied for Load Cases 2, 3, and 4 as shown in **Figure G3–35**.

For Load Case 2, there are two loading conditions shown; both of them have to be checked independently. The eccentricities are calculated as follows:

$E_x = 0.15\, B_x = 0.15\,(170) = 25.5$ ft, and

$E_y = 0.15\, B_y = 0.15\,(240) = 36$ ft

3.9.8 Design Pressures for C&C

Design pressure for C&C is obtained by the following equation.

$$p = q_h[(GC_p) - (GC_{pi})] \qquad \text{(Eq. 6-22)}$$

where

q_h = 13.2 psf for Case 1

(GC_p) = External pressure coefficient (see Figure 6-11 of the Standard)

(GC_{pi}) = ±0.18, the internal pressure coefficient for enclosed buildings

Wall Design Pressures

The pressure coefficients (GC_p) are a function of effective wind area (**Table G3–51**). The definition of effective wind area for a C&C panel is the span length multiplied by an effective width that need not be less than one-third the span length (see Section 6.2 of the Standard). The effective wind areas, A, for wall components are as follows.

Window Unit

$A = 3(4) = 12$ ft^2 (controls)

Wall Stud

larger of $\quad A = 10(1.33) = 13.3$ ft^2

or $\quad\quad\;\; A = 10(10/3) = 33.3$ ft^2 (controls)

Width of Corner Zone 5

smaller of $\quad a = 0.1(170) = 17$ ft

or $\quad\quad\quad\;\; a = 0.1(240) = 24$ ft

or $\quad\quad\quad\;\; a = 0.4(37.3) = 14.9$ ft (controls)

but not less than the smaller of

$\quad\quad\quad\quad\quad\; a = 0.04(170) = 6.8$ ft

or $\quad\quad\quad\;\; a = 0.04(240) = 9.6$ ft

and not less than $\;\; a = 3$ ft

Table G3–47 External Pressures for Wind Normal to Wall W2

Surface type	Surface designation	z or x (ft)	q (psf)	C_p	External pressure (psf)	Design pressures (psf) $(+GC_{pi})$	Design pressures (psf) $(-GC_{pi})$
Walls	W2, W6	0 to 15	10.1	+0.80	+6.9	+4.5	+9.3
		20	10.9	+0.80	+7.4	+5.0	+9.8
		30	12.3	+0.80	+8.4	+6.0	+10.8
	W4, W8	0 to 30	13.2	−0.42	−4.7	−7.1	−2.3
	W1, W3, W5, W7	0 to 30	13.2	−0.70	−7.9	−10.3	−5.5
Roofs (⊥ to ridge)	A1, C2		13.2	−0.25	−2.8	−5.2	−0.4
			13.2	+0.25	+2.8	+0.4	+5.2
	A2, C1		13.2	−0.60	−6.7	−9.1	−4.3
Roofs (∥ to ridge)	A3, C3	0 to 37.3	13.2	−0.90	−10.1	−12.5	−7.7
		37.3 to 70	13.2	−0.50	−5.6	−8.0	−3.2
	B1 & B2	0 to 37.3	13.2	−0.90	−10.1	−12.5	−7.7
		37.2 to 74.6	13.2	−0.50	−5.6	−8.0	−3.2
		74.6 to 240	13.2	−0.30	−3.4	−5.8	−1.0

Note: $q_h = 13.2$ psf; $G = 0.85$.

Table G3–48 External Pressures for Wind Normal to Wall W4

Surface type	Surface designation	z or x (ft)	q (psf)	C_p	External pressure (psf)	Design pressures (psf) $(+GC_{pi})$	Design pressures (psf) $(-GC_{pi})$
Walls	W4, W8	0 to 15	10.1	+0.80	+6.9	+4.5	+9.3
		20	10.9	+0.80	+7.4	+5.0	+9.8
		30	12.3	+0.80	+8.4	+6.0	+10.8
	W6, W2	0 to 30	13.2	−0.42	−4.7	−7.1	−2.3
	W1, W3, W5, W7	0 to 30	13.2	−0.70	−7.9	−10.3	−5.5
Roofs (⊥ to ridge)	C1, A2		13.2	−0.25	−2.8	−5.2	−0.4
			13.2	+0.25	+2.8	+0.4	+5.2
	C2, A1		13.2	−0.60	−6.7	−9.1	−4.3
Roofs (∥ to ridge)	A3, C3	0 to 37.3	13.2	−0.90	−10.1	−12.5	−7.7
		37.3 to 70	13.2	−0.50	−5.6	−8.0	−3.2
	B1, B2	0 to 37.3	13.2	−0.90	−10.1	−12.5	−7.7
		37.2 to 74.6	13.2	−0.50	−5.6	−8.0	−3.2
		74.6 to 240	13.2	−0.30	−3.4	−5.8	−1.0

Note: $q_h = 13.2$ psf; $G = 0.85$.

Table G3-49 External Pressures for Wind Normal to Wall W3

Surface type	Surface designation	z (ft)	q (psf)	C_p	External pressure (psf)	Design pressures (psf)	
						$(+GC_{pi})$	$(-GC_{pi})$
Walls	W3	0 to 15	10.1	+0.80	+6.9	+4.5	+9.3
		20	10.9	+0.80	+7.4	+5.0	+9.8
		30	12.3	+0.80	+8.4	+6.0	+10.8
	W1, W7, W5	0 to 30	13.2	−0.50	−5.6	−8.0	−3.2
	W2, W4, W6, W8	0 to 30	13.2	−0.70	−7.9	−10.3	−5.5
Roofs (⊥ to ridge)	B1		13.2	−0.25	−2.8	−5.2	−0.4
			13.2	+0.25	+2.8	+0.4	+5.2
	A3, B2, C3		13.2	−0.60	−6.7	−9.1	−4.3
Roofs (∥ to ridge)	A1, A2, C1, C2	0 to 37.3	13.2	−0.90	−10.1	−12.5	−7.7
		37.3 to 74.6	13.2	−0.50	−5.6	−8.0	−3.2
		74.6 to 170	13.2	−0.30	−3.4	−5.8	−1.0

Note: q_h = 13.2 psf; G = 0.85.

Table G3-50 External Pressures for Wind Normal to Wall W1-W7-W5

Surface type	Surface designation	z (ft)	q (psf)	C_p	External pressure (psf)	Design pressures (psf)	
						$(+GC_{pi})$	$(-GC_{pi})$
Walls	W1, W7, W5	0 to 15	10.1	+0.80	+6.9	+4.5	+9.3
		20	10.9	+0.80	+7.4	+5.0	+9.8
		30	12.3	+0.80	+8.4	+6.0	+10.8
	W3	0 to 30	13.2	−0.50	−5.6	−8.0	−3.2
	W2, W4, W6, W8	0 to 30	13.2	−0.70	−7.9	−10.3	−5.5
Roofs (⊥ to ridge)	A3, B2, C3		13.2	−0.25	−2.8	−5.2	−0.4
			13.2	+0.25	+2.8	+0.4	+5.2
	B1		13.2	−0.60	−6.7	−9.1	−4.3
Roofs (∥ to ridge)	A1, A2, C1, C2	0 to 37.3	13.2	−0.90	−10.1	−12.5	−7.7
		37.3 to 74.6	13.2	−0.50	−5.6	−8.0	−3.2
		74.6 to 170	13.2	−0.30	−3.4	−5.8	−1.0

Note: q_h = 13.2 psf; G = 0.85.

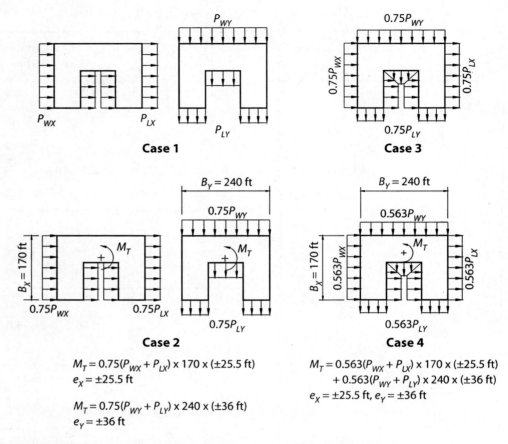

Figure G3–35   Design Wind Load Cases for Wind Normal to Walls W2 and W3

Table G3–51   Wall (GC_p) for Ex. 9, by Zone

Component	A (ft^2)	Zones 4 and 5 ($+GC_p$)	Zone 4 ($-GC_p$)	Zone 5 ($-GC_p$)
Window	12	+0.99	−1.09	−1.37
Wall Stud	33.3	+0.91	−1.01	−1.22

Table G3–52   Controlling Design Pressures for Wall Components (psf), by Zone

Component	Zone 4 Positive	Zone 4 Negative	Zone 5 Positive	Zone 5 Negative
Window unit	+15.4	−16.8	+15.4	−20.5
Mullion	+14.4	−15.7	+14.4	−18.5

Typical Design Pressure Calculations

Controlling design pressures for wall components are presented in **Table G3–52**.

Controlling negative design pressure for window unit in Zone 4 of walls

$= 13.2[(-1.09) - (\pm 0.18)]$
$= -16.8$ psf (positive internal pressure controls)

Controlling positive design pressure for window unit in Zone 4 of walls

$= 13.2[(+0.99) - (\pm 0.18)]$
$= 15.4$ psf (negative internal pressure controls)

The design pressures are the algebraic sum of external and internal pressures. Controlling negative pressure is obtained with positive internal pressure, and controlling positive pressure is obtained with negative internal pressure.

The edge zones for the walls are arranged at exterior corners, as shown in **Figure G3–36**.

Roof Design Pressures

The C&C roof pressure coefficients are given in Figure 6-11C of the Standard and presented in **Tables G3–53** and **G3–54**. The pressure coefficients are a function of the effective wind area. The definition of effective wind area for a component or cladding panel is the span length multiplied by an effective

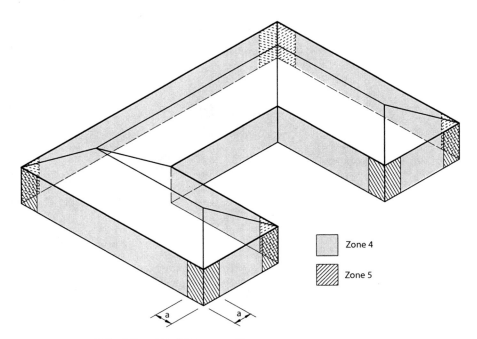

Figure G3–36   Component and Cladding Wall Pressure Zones

Table G3–53 Roof External Pressure Coefficients (GC_p), by Zone

	Positive	Negative	
	Zones 1, 2	Zone 1	Zone 2
A (ft^2)	GC_p	GC_p	$-GC_p$
1633	+0.3	−0.80	−1.2

Table G3–54 Roof Design Pressures (psf), by Zone

Component	Zones 1, 2	Zone 1	Zone 2
	Positive	Negative	Negative
Roof rafter	+6.3	−12.9	−18.2

width that need not be less than one-third the span length (see Section 6.2 of the Standard). The effective wind areas, A, for the roof trusses are as follows.

Roof Truss Top Chord

larger of $A = 70(2.0) = 140$ ft^2
or $A = 70(70/3) = 1633$ ft^2 (controls)

Note 7 of Figure 6-11C of the Standard says that for hip roofs with $\theta \leq 25°$, Zone 3 may be treated as Zone 2.

The design pressures are the algebraic sum of external and internal pressures. Controlling negative pressure is obtained with positive internal pressure, and controlling positive pressure is obtained with negative internal pressure.

The edge zones for the hip roof are arranged as shown in **Figure G3–37**.

3.10 Example 10: 50-ft × 20-ft Billboard Sign on Poles (Flexible) 60 ft Above Ground

In this example, design wind-forces for a tall billboard solid sign are determined. The example illustrates two items: (1) determination of G_f for a flexible structure, and (2) use of force coefficient from Figure 6-20 for solid signs and Figure 6-21 for other structures. The dimensions of the billboard sign are shown in **Figure G3–38**. The billboard sign data are as shown below.

Location Interstate highway in Iowa

Terrain Flat and open terrain

Dimensions 50-ft × 20-ft sign mounted on two 16-in.-diameter steel pipe supports; bottom of the sign is 60 ft above ground

Structural Characteristics Tall flexible structure; estimated fundamental frequency is 0.7 Hz and critical damping ratio is 0.01
(The natural frequency of a structure can be calculated in different ways. It has been predetermined for this example.)

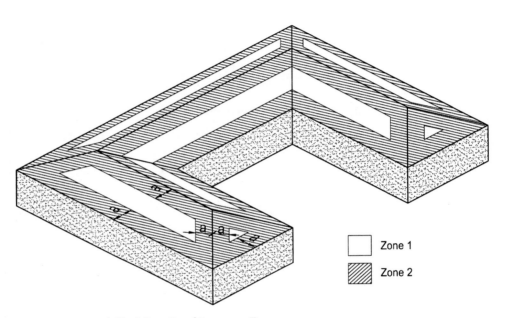

Figure G3–37   Component and Cladding Roof Pressure Zones

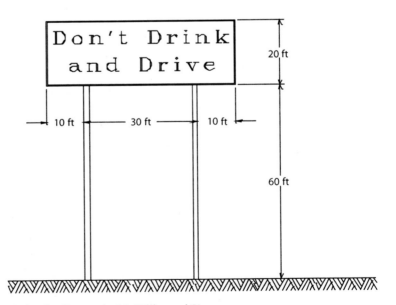

Figure G3–38   Characteristics for Example 10, Billboard Sign

3.10.1 Exposure and Building Classification

The sign is located in an open area. It does not fit Exposures B or D; therefore, Exposure C is used (see Sections 6.5.6.2 and 6.5.6.3 of the Standard).

Failure of the sign represents low hazard to human life since it is located away from the highway and is not in a populated area. The structure can be classified as Category I (see Table 1-1).

3.10.2 Basic Wind Speed

The wind speed map (Figure 6-1 of the Standard) has only one value of wind speed in the middle of the country. Exact location of the sign in Iowa is not important. The basic wind speed $V = 90$ mph.

3.10.3 Velocity Pressures

The velocity pressures are computed using

$$q_z = 0.00256 \, K_z K_{zt} K_d V^2 I \text{ psf} \qquad \text{(Eq. 6-15)}$$

where

V = 90 mph

I = 0.87 for Category I (see Table 6-1 of the Standard)

K_{zt} = 1.0 because of flat terrain

K_d = 0.85 for solid sign (see Table 6-4 of the Standard)

K_z = Values from Table 6-3 of the Standard for z of 30, 60, and 80 ft. More divisions of z are not justified because loads on pipe supports are small compared to the ones on the sign.

Values for K_z and the resulting velocity pressures are given in **Table G3–55**.

3.10.4 Design Force for the Structure

The design force for the Solid Sign (Section 6.5.14) is given by

$$F = q_h \, G \, C_f A_s \qquad \text{(Eq. 6-27)}$$

The design force for the support poles (Section 6.5.15) is given by

$$F = q_z \, G \, C_f A_f \qquad \text{(Eq. 6-28)}$$

where

q_z = Value as determined previously

q_h = Value determined at the top of the sign (see Figure 6-20)

G = Gust effect factor to be calculated by Eq. 6-8 because $f < 1$ Hz.

A_s = Gross area of solid sign; $50 \times 20 = 1,000$ ft^2

A_f = Area projected normal to wind; 1.33 ft/ft of support height

C_f = Force coefficient values from Figures 6-20 and 6-21 of the Standard

Table G3–55   Velocity Pressures (psf)

Height (ft)	K_z	q_z (psf)
30	0.98	15.0
60	1.13	17.3
80	1.21	18.6

Gust Effect Factor

The gust effect factor is determined as

$$G = 0.925 \left[\frac{1 + 1.7 I_{\bar{z}} \sqrt{g_Q^2 Q^2 + g_R^2 R^2}}{1 + 1.7 g_V I_{\bar{z}}} \right] \quad \text{(Eq. 6-8)}$$

where

$\quad I_{\bar{z}}$ = Value from Eq. 6-5 of the Standard

$\quad g_Q, g_V$ = Value taken as 3.4 (see Section 6.5.8.2 of the Standard)

$\quad g_R$ = Value from Eq. 6-9 of the Standard

$\quad Q$ = Value determined from Eq. 6-6 of the Standard

$\quad R$ = Value determined from Eq. 6-1 of the Standard

$\quad I_{\bar{z}}$ = Equivalent height of the structure, it is used to determine nominal value of $I_{\bar{z}}$; for buildings, the recommended value is $0.6h$, but for the sign, it is the middle of the billboard area or 70 ft

$\quad c, l, \bar{\epsilon}$ = Value given in Table 6-2 of the Standard

$$I_{\bar{z}} = c \left(\frac{33}{\bar{z}} \right)^{1/6} = 0.2 \left(\frac{33}{70} \right)^{1/6} = 0.176 \quad \text{(Eq. 6-5)}$$

$$L_{\bar{z}} = l \left(\frac{\bar{z}}{33} \right)^{\bar{\epsilon}} = 500 \left(\frac{70}{33} \right)^{1/5} = 581 \text{ ft} \quad \text{(Eq. 6-7)}$$

$$Q^2 = \frac{1}{1 + 0.63 \left[\frac{B+h}{L_{\bar{z}}} \right]^{0.63}} \quad \text{(Eq. 6-6)}$$

$$= \frac{1}{1 + 0.63 \left[\frac{50+20}{581} \right]^{0.63}} = 0.858$$

Note: In Eq. 6-6, B and h are the dimensions of the sign.

$$\bar{V}_{\bar{z}} = \bar{b}\left(\frac{\bar{z}}{33}\right)^{\bar{\alpha}} V\left(\frac{88}{60}\right) = 0.65\left(\frac{70}{33}\right)^{1/6.5}(90)\left(\frac{88}{60}\right) = 96.3 \qquad \text{(Eq. 6-14)}$$

Note: V is the basic (3-s gust) wind speed in mph.

$$N_1 = \frac{n_1 L_{\bar{z}}}{\bar{V}_{\bar{z}}} = \frac{(0.7)(581)}{96.3} = 4.22 \qquad \text{(Eq. 6-12)}$$

Note: n_1 is the fundamental frequency of the structure.

$$R_n = \frac{7.47 N_1}{(1 + 10.3 N_1)^{5/3}} = 0.0564 \qquad \text{(Eq. 6-11)}$$

For R_h,

$$\eta = \frac{4.6 n_1 h}{\bar{V}_{\bar{z}}} = \frac{(4.6)(0.7)(80)}{96.3} = 2.675$$

$$R_h = \frac{1}{\eta} - \frac{1}{2\eta^2}\left(1 - e^{-2\eta}\right) = 0.3043 \qquad \text{(Eq. 6-13a)}$$

Note: h is taken as 80 ft because resonance response depends on full height.

For R_B (assuming $B = 50$ ft),

$$\eta = \frac{4.6 n_1 B}{\bar{V}_{\bar{z}}} = \frac{(4.6)(0.7)(50)}{96.3} = 1.672 \qquad \text{(Eq. 6-13b)}$$

$$R_B = \frac{1}{\eta} - \frac{1}{2\eta^2}\left(1 - e^{-2\eta}\right) = 0.4255$$

For R_L (assuming depth $L = 2$ ft),

$$\eta = \frac{15.4 \, n_1 L}{\bar{V}_{\bar{z}}} = \frac{(15.4)(0.7)(2)}{96.3} = 0.2239 \qquad \text{(Eq. 6-13c)}$$

$$R_L = \frac{1}{\eta} - \frac{1}{2\eta^2}\left(1 - e^{-2\eta}\right) = 0.8661$$

$$g_R = \sqrt{2\ln(3{,}600 n_1)} + \frac{0.577}{\sqrt{2\ln(3{,}600 n_1)}} \qquad \text{(Eq. 6-9)}$$

$$g_R = 4.1$$

$$R^2 = \frac{1}{\beta} R_n R_h R_B (0.53 + 0.47 R_L)$$

$$= \frac{1}{0.01}(0.0564)(0.3043)(0.4255)\left[0.53 + (0.47)(0.8661)\right]$$

$$R^2 = 0.684 \quad \text{(Eq. 6-10)}$$

$$G_f = 0.925\left[\frac{1 + 1.7 I_{\bar{z}}\sqrt{g_Q^2 Q^2 + g_R^2 R^2}}{1 + 1.7 g_v I_{\bar{z}}}\right]$$

$$= 0.925\left[\frac{1 + 1.7(0.176)\sqrt{(3.4)^2(0.858) + (4.1)^2(0.684)}}{1 + 1.7(3.4)(0.176)}\right]$$

$$= 1.093 \quad \text{(Eq. 6-8)}$$

Force Coefficient (C_f) for Supports

The supports are round. From Figure 6-21 of the Standard,

$$D\sqrt{q_z} = 1.33\sqrt{15.0} = 5.2 > 2.5 \text{ and}$$
$$h/D = 60/1.33 = 45$$

For moderately smooth surface,

$$C_f = 0.7$$

Design Force on Supports

Force, $F = q_z G_f C_f A_f$

For one pole

0 to 30 ft	$F = 15.0(1.093)(0.7)(1.33) = 15.3$ plf
30 to 60 ft	$F = 17.3(1.093)(0.7)(1.33) = 17.7$ plf

For two poles

0 to 30 ft	$F = 30.6$ plf
30 to 60 ft	$F = 35.4$ plf

For a 1-ft horizontal strip of the sign

$$F = 18.6(1.093)(1.2)(50) = 1{,}220 \text{ plf}$$

The force on the sign follows three cases (see **Figure G3–39**).

Force Coefficients for Solid Sign

$$B/s = 50/20 = 2.5 \text{ (Aspect ratio in Figure 6-20)}$$
$$s/h = 20/80 = 0.25 \text{ (Clearance ratio in Figure 6-20)}$$

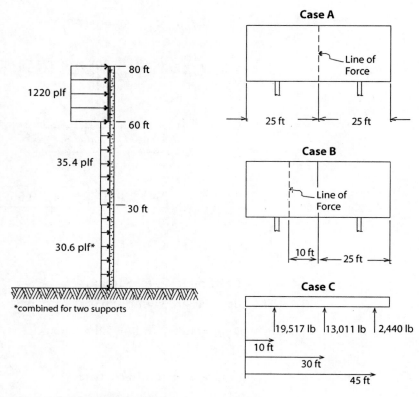

Figure G3–39   Design Forces for the Billboard Sign

C_f = 1.8 for Case A and Case B (interpolation in Figure 6-20)

C_f = 2.4 (0 to 20 ft); 1.6 (20 to 40 ft); 0.6 (40 to 50 ft) for Case C (linear interpolation)

Design force on sign (see Figure 6-20 for Cases)

Case A: F = 18.6 (1.093) (1.8) (1000) = 36,594 lbs (or 1830 plf of sign height)

Case B: Eccentricity e = 0.2 (50) = 10 ft

Force of 36,594 lbs acts at 10 ft from the center of the sign

Case C: F = 18.6 (1.093) (2.4) (400) = 19,517 lbs acts at 10 ft from an edge of sign

F = 18.6 (1.093) (1.6) (400) = 13,011 lbs acts at 30 ft from the edge of sign

F = 18.6 (1.093) (0.6) (200) = 2,440 lbs acts at 45 ft from the edge of sign

Note: Total force for Cases A and C are comparable; 36,594 vs 34,968 lbs. Eccentricity for Case B is higher than that of Case C. Case C loading may be critical for the design of sign structure.

3.10.5 Limitation

In certain circumstances for circular members, across-wind response due to vortex shedding can be critical. The Standard does not provide a procedure to assess across-wind response, but suggests obtaining guidance from recognized literature (see Section 6.5.2 of the Standard).

3.10.6 Force on Components and Cladding

Eq. 6-25 of the Standard is

$$F = q_z G C_f A_f$$

The values of q_z are the same as MWFRS, except the value of $G = 0.85$. The design forces can be determined using appropriate C_f and A_f for each component or cladding panel.

3.11 Example 11: Domed Roof Building

Figure G3–40 illustrates the domed roof building used for a church in this example. Building data are as shown here.

Location Baton Rouge, Louisiana

Topography Homogeneous

Terrain Open

Dimensions 100 ft diameter in plan
Eave height of 20 ft
Dome roof height of 50 ft

Framing Steel framed dome roof
Metal deck roofing

Cladding Location is outside a wind-borne debris region, so no glazing protection is required

Domed roofs are outside the scope of Method 1, Simplified Procedure, of ASCE 7-05. Method 2, Analytical Procedure, is used. The building is less than 60 feet tall, so it is possible to use low-rise provisions of Section 6.5.12.2.2. However, because dome-shaped roofs are not specifically covered, the adaptation of the low-rise "pseudo pressure" coefficients to buildings are outside the scope of the research and is not recommended. Therefore, the "all heights method" of Section 6.5.12.2.1 of the Standard is used.

3.11.1 Exposure

The building is located in an open terrain area; according to Section 6.5.6.3 of the Standard, Exposure C is used.

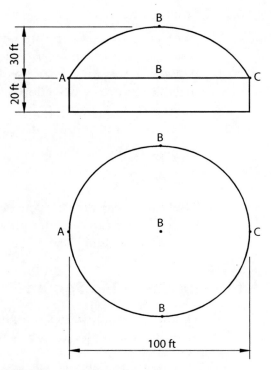

Figure G3–40   Characteristics for Example 11, Domed Roof Building

3.11.2   Building Classification

The building is a church, so it will have more than 300 people congregating in one area. Therefore, building Category III is appropriate (see Table 1-1 of the Standard).

3.11.3   Enclosure

The building is designed to be enclosed. It is not located within a wind-borne debris region, so glazing protection is not required.

3.11.4   Basic Wind Speed

Selection of the basic wind speed is addressed in Section 6.5.4 of the Standard. Baton Rouge, Louisiana, is located halfway between the 100-mph and 110-mph contours; therefore, the basic wind speed $V = 105$ mph (see Figure 6-1b of the Standards).

3.11.5   Velocity Pressures

The velocity pressures are computed using

$$q_z = 0.00256 K_z K_{zt} K_d V^2 I \text{ psf} \tag{Eq. 6-15}$$

112 Wind Loads: Guide to the Wind Load Provisions of ASCE 7-05

where

K_z = Value obtained from Table 6-3 of the Standard: For Exposure C, values for Cases 1 and 3 are the same

K_{zt} = 1.0 for homogeneous topography

K_d = 0.95 for round tanks and similar structures (see Table 6-4 of the Standard)

V = 105 mph

I = 1.15 for Category III classification (see Table 6-1 of the Standard)

therefore,

$$q_z = 0.00256 K_z (1.0)(0.95)(105)^2(1.15) = 30.8 K_z \text{ psf}$$

Values for K_z and the resulting velocity pressures are given in **Table G3–56**. Wall pressures will be evaluated at mid-height = 20 ft/2 = 10 ft /15 ft; use q_z at 15 ft.

Design Wind Pressures for the MWFRS

Wall Pressures

The walls of a round building are not specifically covered by the Standard. The values for the force coefficients for round tanks and chimneys from Figure 6-19 is used to determine the effect of the wall pressures on the MWFRS. The values of the force coefficients for round tanks vary with the aspect ratio of height to diameter and with the surface roughness.

The value of q_z varies from 26.2 psf at the ground to 27.7 psf at the eave line. Therefore, the ratio $D\sqrt{q_z}$ varies from $100\sqrt{26.2} = 512$ to $100\sqrt{27.7} = 527$ both of which are much greater than 2.5; therefore, the first set of values for C_f for round tanks in Figure 6-21 of the Standard is used. Any projections on the exterior skin of the building are assumed to be less than 2 ft; therefore, D'/D would be less than 2 ft/100 ft = 0.02, so the building is considered,

Table G3–56 q_z Velocity Pressures

Height (ft)	MWFRS		C&C	
	K_z	q_z (psf)	K_z	q_z (psf)
0–15	0.85	26.2	0.85	26.2
Eave height = 20	0.90	27.7	0.90	27.7
Top of dome = 50	1.09	33.6	1.09	33.6

moderately smooth. The height of the entire structure (h = 50 ft) is used for the aspect ratio, since the wind has to travel over the dome. Therefore, h/D = 50 ft/100 ft = 0.5, which is less than 1, resulting in C_f = 0.5.

The force on the walls represents the total drag of the wind on the walls of the building, both windward and leeward. Since it is not the typical pressures applied normal to the wall surfaces, ignore internal pressures, as they cancel out in the net drag calculation.

$$\text{Total drag force on walls} = F = q_z G C_f A_f \qquad \text{(Eq. 6-25)}$$

where

q_z = q at the centroid of A_f—centroid of A_f is at wall mid-height = 20 ft/2 = 10 ft

q = 26.2 psf (at 10 ft)

G = 0.85, the gust effect factor for rigid structures

C_f = 0.5

A_f = 100 ft × 20 ft = 2,000 sf

therefore,

$$F = 26.2(0.85)(0.5)(2,000) = 22,270 \text{ lb}$$

Domed Roof Pressures

The roof pressure coefficients for a domed roof (**Table G3–57**) are taken from Figure 6-7 of the Standard. The height from the ground to the spring line of the dome, h_D = 20 ft. The height of the dome itself from the spring line to the top of the dome, f = 30 ft. Determine C_p for a rise to diameter ratio, f/D = 30/100 = 0.30; and a base height to diameter ratio, h_D/D = 20/100 = 0.20. Interpolation from Figure 6-7 of the Standard is required.

Two load cases are required for the MWFRS loads on domes: Cases A and B. Case A is based on linear interpolation of C_p values from point A to B and from point B to C (see **Figure G3–40** for the locations of points A, B, and C). Case B uses the pressure coefficient at A for the entire front area of the dome up to an angle θ = 25°, then interpolates the values for the rest of the dome as in Case A.

Table G3–57 Domed Roof C_p (at f/D = 0.30)

Point on dome	$h_D/D = 0$	$h_D/D = 0.20$	$h_D/D = 0.25$	$h_D/D = 0.50$
A	+0.5	−0.04	−0.18	—
B	−0.78	−0.97	—	−1.26
C	0	−0.20	—	−0.50

Case A

For design purposes, interpolate the pressure coefficients at points at 10-ft intervals along the dome. Values of pressure coefficients C_p are shown in **Table G3–58**.

Case B

Determine the point on the front of the dome at which $\theta = 25°$. The point is 36.2 ft from the center of the dome, therefore 13.8 ft from point A. The pressure coefficient at A shall be used for the section from A to an arc 13.8 ft from A. The remainder of the dome pressures are based on linear interpolation between the 25° point and point B; and then from point B to C. Values of pressure coefficients C_p are shown in **Table G3–59**.

Internal Pressure Coefficient for Domed Roof

The building is not in a wind-borne debris region, so glazing protection is not required. The building is assumed to be an enclosed building.

The net pressure on any surface is the difference in the external and internal pressures on the opposites sides of that surface.

$$p = qGC_p - q_i(GC_{pi})$$ (Eq. 6-17)

For enclosed buildings

$$GC_{pi} = \pm 0.18$$ (Figure 6-5)

q_i is taken as $q_{(hD+f)} = 33.6$ psf

Design internal pressure

$$q_i(GC_{pi}) = 33.6\ (\pm 0.18) = \pm 6.1\ \text{psf}$$

Table G3-58 Interpolated Domed Roof C_p (Case A)

Segment	Start point	+10 ft	+20 ft	+30 ft	+40 ft	End point
A to B	−0.04	−0.23	−0.41	−0.60	−0.78	−0.97
B to C	−0.97	−0.82	−0.66	−0.51	−0.35	−0.20

Table G3-59 Interpolated Domed Roof C_p (Case B)

Segment	Start point	+10 ft	+13.8 ft	+20 ft	+30 ft	+40 ft	End point
A to B	−0.04		−0.04	−0.20	−0.46	−0.71	−0.97
B to C	−0.97	−0.82		−0.66	−0.51	−0.35	−0.20

Design Wind Pressures for Domed Roof

The design pressures for this building (**Figure G3–41**) are obtained by the equation

$$p = qGC_p - q_i(GC_{pi}) \qquad \text{(Eq. 6-17)}$$

where

$$q = q_{(hD+f)} = 33.6 \text{ psf}$$
$$G = 0.85, \text{ the gust effect factor for rigid buildings and structures}$$
$$C_p = \text{External pressure coefficient}$$
$$q_i = q_h \text{ for all surfaces since the building is enclosed}$$
$$GC_{pi} = \pm 0.18, \text{ the internal pressure coefficient for enclosed buildings}$$

therefore

$$p = 33.6(0.85)C_p - 33.6(\pm 0.18) = 28.6C_p \pm 6.1.$$

Values of design pressures for MWFRS are shown in **Table G3–60**.

3.11.6 Design Wind Load Cases

Section 6.5.12.3 of the Standard requires that any building whose wind loads have been determined under the provisions of Sections 6.5.12.2.1 and 6.5.12.2.3 shall be designed for wind load cases as defined in Figure 6-9. However, since the building is round, the cases as shown do not apply. There is a possibility of non-symmetrical action by the wind, causing some torsion. Load Case 2, with the reduced calculated horizontal load and moment using eccentricity of 15 ft, could be applied to the cylindrical wall portion of the building.

3.11.7 Design Pressures for C&C

Design pressure for components and cladding (**Figure G3–42**) is obtained by

$$p = q_h[(GC_p) - (GC_{pi})], \qquad \text{(Eq. 6-22)}$$

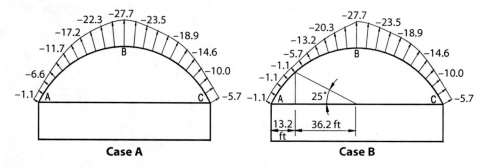

Figure G3–41   *MWFRS External Pressures. Note: Internal pressure of ±6.1 psf to be added.*

Table G3–60   Domed Roof Design Pressures for MWFRS (psf)

Surface of Domed Roof	Location (ft)	C_p	External pressure (psf)	Design pressures (psf) $(+GC_{pi})$	Design pressures (psf) $(-GC_{pi})$
Case A	Point A – 0 ft	–0.04	–1.1	–7.2	5.0
	10	–0.23	–6.6	–12.7	–0.5
	20	–0.41	–11.7	–17.8	–5.6
	30	–0.60	–17.2	–23.3	–11.1
	40	–0.78	–22.3	–28.4	–16.2
	Point B – 50 ft	–0.97	–27.7	–33.8	–21.6
	60	–0.82	–23.5	–29.6	–17.4
	70	–0.66	–18.9	–25.0	–12.8
	80	–0.51	–14.6	–20.7	–8.5
	90	–0.35	–10.0	–16.1	–3.9
	Point C – 100 ft	–0.20	–5.7	–11.8	0.4
Case B	Point A – 0 ft	–0.04	–1.1	–7.2	5.0
	θ = 25°; 13.8 ft	–0.04	–1.1	–7.2	5.0
	20	–0.20	–5.7	–11.8	0.4
	30	–0.46	–13.2	–19.3	–7.1
	40	–0.71	–20.3	–26.4	–14.2
	Point B – 50 ft	–0.97	–27.7	–33.8	–21.6
	60	–0.82	–23.5	–29.6	–17.4
	70	–0.66	–18.9	–25.0	–12.8
	80	–0.51	–14.6	–20.7	–8.5
	90	–0.35	–10.0	–16.1	–3.9
	Point C – 100 ft	–0.20	–5.7	–11.8	0.4

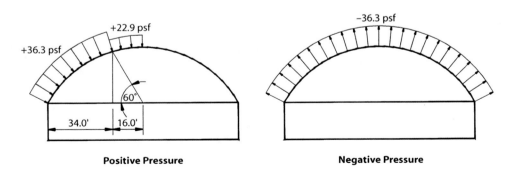

Figure G3–42   Component Design Pressures

where

$q_h = q_{(hD+f)}$ = 33.6 psf for all domed roofs calculated at height h_D+f

$q_i = q_{(hD+f)}$ = 33.6 psf for positive and negative internal pressure

(GC_p) = External pressure coefficient (see Figure 6-16 of the Standard)

(GC_{pi}) = ±0.18 for internal pressure coefficient (see Figure 6-5 of the Standard)

Wall Design Pressures

The Standard does not address component and cladding wall loads for round buildings. The designer might consider segmenting these walls into small widths and using the roof coefficients for the walls. The designer could also use the Figures 6-11A or 6-17 (depending on h), and apply the coefficients for wall zone 4. These values will come close to the domed roof coefficients of ± 0.9.

Domed Roof Design Pressures

The C&C domed roof pressure coefficients (**Table G3–61**) are from Figure 6-16 of the Standard. This figure is valid only for domes of certain geometric parameters. The base height to diameter ratio, $h_D/D = 20/100 = 0.20$, which is in the range of 0 to 0.5 for Figure 6-16. The rise to diameter ratio, $f/D = 30/100 = 0.30$ which is in the range of 0.2 to 0.5 for Figure 6-16. Therefore, it is valid to use Figure 6-16 for this dome.

The design pressures are the algebraic sum of external and internal pressures. Positive internal pressure provides controlling negative pressures, and negative internal pressure provides the controlling positive pressure. These design pressures act across the roof surface (interior to exterior).

$$p = qGC_p - q_i(GC_{pi})$$

$$p = 33.6GC_p - 33.6(\pm 0.18) = 33.6GC_p \pm 6.1$$

Design pressures are summarized in **Table G3–62**.

These pressures are for the front half of the dome. The back half would experience only the negative value of −36.3 psf. However, since all wind directions must be taken into account, and since each element would at some point be considered to be in the front half of the dome, each element must be designed for both positive and negative values.

Table G3–61 Roof External Pressure Coefficient (GC_p)

Zone	Positive	Negative
0° to 60°	+0.9	−0.9
60° to 90°	+0.5	−0.9

Note: Coefficients are from Figure 6–16 of the Standard.

Table G3–62 Roof Design Pressures (psf)

Zone	Positive	Negative
0° to 60°	+36.3	−36.3
60° to 90°	+22.9	−36.3

3.12 Example 12: Unusually Shaped Building

This example demonstrates calculation of wind loads for an unusually shaped building, as shown in **Figure G3–43.** Building data are as shown here.

Location San Francisco, California

Topography Homogeneous

Terrain Suburban

Dimensions 100-ft × 100-ft overall in plan with a 70-ft × 70-ft wedge cut off
Flat roof with eave height of 15 ft

Framing Steel joist, beam, column roof framing with X-bracing.

Cladding Location is outside a wind-borne debris region, so no glazing protection is required.

Non-symmetrical buildings are outside the scope of Method 1, Simplified Procedure, of ASCE 7-05. Therefore, Method 2, Analytical Procedure, is used. The building is less than 60 ft tall, so it is possible to use low-rise provisions of Section 6.5.12.2.2. However, because unusually shaped buildings are not specifically covered, the adaptation of the low-rise "pseudo pressure" coefficients to buildings outside the scope of the research is not recommended. Therefore, the "all heights method" of Section 6.5.12.2.1 is used.

3.12.1 Exposure

The building is located in a suburban terrain area; according to Section 6.5.6.3 of the Standard, Exposure B is used.

3.12.2 Building Classification

The building is an office building. It is not considered an essential facility, nor is it likely to be occupied by 300 persons in a single area at one time. Therefore, building Category II is appropriate (see Table 1-1 of the Standard).

3.12.3 Enclosure

The building is designed to be enclosed. It is not located within a wind-borne debris region, so glazing protection is not required.

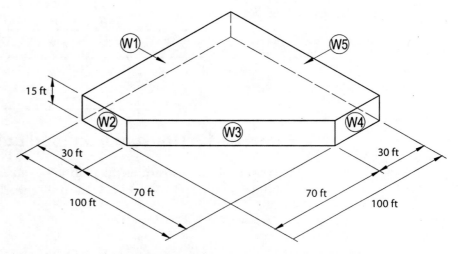

Figure G3–43   Building Characteristics for Example 12, Unusually Shaped Building

3.12.4   Basic Wind Speed

Selection of the basic wind speed is addressed in Section 6.5.4 of the Standard. San Francisco, California, is located in the 85-mph zone; therefore, the basic wind speed $V = 85$ mph (see Figure 6-1 of the Standard).

3.12.5   Velocity Pressures

The velocity pressures are computed using

$$q_z = 0.00256 K_z K_{zt} K_d V^2 I \text{ psf} \qquad \text{(Eq. 6-15)}$$

where

K_z = Value obtained from Table 6-3: Case 1 for C&C and Case 2 for MWFRS

K_{zt} = 1.0 for homogeneous topography

K_d = 0.85 for buildings (see Table 6-4 of the Standard)

V = 85 mph

I = 1.0 for Category II classification (see Table 6-1 of the Standard)

therefore,

$$q_z = 0.00256 K_z (1.0)(0.85)(85)^2(1.0) = 15.72 K_z \text{ psf}$$

The mean roof height for a flat roof is the eave height $h = 15$ ft. Values for K_z and the resulting velocity pressures for MWFRS and C&C are shown in **Table G3–63**.

External Pressure Coefficients

The values for the external pressure coefficients for the various surfaces are obtained from Figure 6-6 of the Standard for each of the surfaces of the

Table G3–63 q_z Velocity Pressure

	MWFRS		C&C	
Height (ft)	K_z	q_z (psf)	K_z	q_z (psf)
0–15	0.57	9.0	0.70	11.0
Eave height = 15	0.57	9.0	0.70	11.0

building shown in **Figure G3–43**. The determination of certain pressure coefficients is based on aspect ratios. The overall dimensions for L and B are used.

$L/B = 100/100 = 1.00$

$h/L = 15/100 = 0.15$

$\theta = 0°$

The windward wall C_p is always 0.8, the side walls are always –0.7, and the leeward wall is –0.5 based on an aspect ratio $L/B = 1.0$.

The roof C_p comes from the "wind parallel to a ridge" portion of Figure 6-6 of the Standard. For these flat roofs, C_p varies with h/L and with distance from the leading edge of the roof. For $h/L = 0.15 \leq 0.5$, $C_p = -0.9, -0.5$, or –0.3, depending on the distance from the leading edge. Figure 6-6 also includes the –0.18 case for all roofs; however, this case causes critical loading when combined with transient loads such as snow load or live load. For brevity, the case is not shown.

External pressure coefficients are summarized in **Tables G3–64** through **G3–67**.

3.12.6   Design Wind Pressures for the MWFRS

The design pressures for this building are obtained by the equation

$$p = qGC_p - q_i(GC_{pi}) \qquad \text{(Eq. 6-17)}$$

where

$q = q_z = 9.0$ psf for windward wall at height $z = 15$ ft and below

$q = q_h = 9.0$ psf for leeward wall, side walls, and roof

$q_i = q_h = 9.0$ psf for all surfaces since the building is enclosed

$G = 0.85$, the gust effect factor for rigid buildings and structures

$C_p =$ External pressure coefficient for each surface, as shown in **Figure G3–43**

$(GC_{pi}) = \pm 0.18$, the internal pressure coefficient for enclosed buildings

Table G3–64　External Pressure Coefficients (C_p) for Wind Normal to Wall W1

Surface type	Surface designation	Surface	Distance from windward edge	L/B or h/L	C_p
Walls	W1	Windward		All	+0.80
	W3, W4	Leeward		1.0	−0.50
	W2, W5	Side		All	−0.70
Roof			0 to h	0.15	−0.90*
			h to $2h$	0.15	−0.50*
			> $2h$	0.15	−0.30*

* The values of smaller uplift pressures ($C_p = -0.18$) on the roof can become critical when wind load is combined with roof live load or snow load; load combinations are given in Sections 2.3 and 2.4 of the Standard. For brevity, loading for this value is not shown here.

Table G3–65　External Pressure Coefficients (C_p) for Wind Normal to Wall W5

Surface type	Surface designation	Surface	Distance from windward edge	L/B or h/L	C_p
Walls	W5	Windward		All	+0.80
	W2, W3	Leeward		1.0	−0.50
	W1, W4	Side		All	−0.70
Roof			0 to h	0.15	−0.90
			h to $2h$	0.15	−0.50
			> $2h$	0.15	−0.30

Table G3–66　External Pressure Coefficients (C_p) for Wind Normal to Wall W4

Surface type	Surface designation	Surface	Distance from windward edge	L/B or h/L	C_p
Walls	W4, W3	Windward		All	+0.80
	W1	Leeward		1.0	−0.50
	W2, W5	Side		All	−0.70
Roof			0 to h	0.15	−0.90
			h to $2h$	0.15	−0.50
			> $2h$	0.15	−0.30

Table G3–67 External Pressure Coefficients (C_p) for Wind Normal to Wall W2

Surface type	Surface designation	Surface	Distance from windward edge	L/B or h/L	C_p
Walls	W2, W3	Windward		All	+0.80
	W5	Leeward		1.0	−0.50
	W1, W4	Side		All	−0.70
Roof			0 to h	0.15	−0.90
			h to 2h	0.15	−0.50
			> 2h	0.15	−0.30

For windward walls

$$p = q_z G C_p - q_h(GC_{pi}) = 9.0(0.85)C_p - 9.0(\pm 0.18) = 7.7 C_p \pm 1.6$$

For all other surfaces

$$p = q_h G C_p - q_h(GC_{pi}) = 9.0(0.85)C_p - 9.0(\pm 0.18) = 7.7 C_p \pm 1.6$$

Design pressures are summarized in **Tables G3–68** through **G3–71**. The external roof pressures and their prescribed zones are shown in **Figure G3–44**.

Minimum Design Wind Pressures

Section 6.1.4.1 of the Standard requires that the MWFRS be designed for not less than 10 psf applied to the projection of the building in each orthogonal direction on a vertical plane. This is checked as a separate load case. The application of this load is shown in **Figure G3–45**.

3.12.7 Design Wind Load Cases

Section 6.5.12.3 of the Standard requires that any building whose wind loads have been determined under the provisions of Sections 6.5.12.2.1 and 6.5.12.2.3 shall be designed for wind load cases as defined in Figure 6-9. There are several exceptions noted that require only the use of Load Case 1, the full orthogonal wind case, and Load Case 3, the diagonal wind case approximated by applying 75% of the loads to adjacent faces simultaneously. The exceptions are building types that are not sensitive to torsional wind effects, which are created by Load Cases 2 and 4. One of these exceptions is for one-story buildings less than 30 ft in height, so this example meets that exception and is required only to meet Load Cases 1 and 3. Load Case 1 is calculated above and shown applied in each orthogonal direction in **Figure G3–46**. Load Case 3 is the diagonal wind load case, applied in each of four directions as shown in **Figure G3–47**.

Table G3–68 Design Pressures for Wind Normal to Wall W1

Surface type	Surface designation	z or x (ft)	q (psf)	C_p	External pressure (psf)	Design pressures (psf) $(+GC_{pi})$	$(-GC_{pi})$
Walls	W1	0 to 15	9.0	+0.80	+6.2	+4.6	+7.8
	W3, W4	0 to 15	9.0	−0.50	−3.9	−5.5	−2.3
	W2, W5	0 to 15	9.0	−0.70	−5.4	−7.0	−3.8
Roof		0 to 15	9.0	−0.90	−6.9	−8.5	−5.3
		15 to 30	9.0	−0.50	−3.9	−5.5	−2.3
		30 to 100	9.0	−0.30	−2.3	−3.9	−0.7

Note: q_h = 9.0 psf; G = 0.85.

Table G3–69 Design Pressures for Wind Normal to Wall W5

Surface type	Surface designation	z or x (ft)	q (psf)	C_p	External pressure (psf)	Design pressures (psf) $(+GC_{pi})$	$(-GC_{pi})$
Walls	W5	0 to 15	9.0	+0.80	+6.2	+4.6	+7.8
	W2, W3	0 to 15	9.0	−0.50	−3.9	−5.5	−2.3
	W1, W4	0 to 15	9.0	−0.70	−5.4	−7.0	−3.8
Roof		0 to 15	9.0	−0.90	−6.9	−8.5	−5.3
		15 to 30	9.0	−0.50	−3.9	−5.5	−2.3
		30 to 100	9.0	−0.30	−2.3	−3.9	−0.7

Note: q_h = 9.0 psf; G = 0.85.

Table G3–70 Design Pressures for Wind Normal to Wall W4

Surface type	Surface designation	z or x (ft)	q (psf)	C_p	External pressure (psf)	Design pressures (psf) $(+GC_{pi})$	$(-GC_{pi})$
Walls	W4, W3	0 to 15	9.0	+0.80	+6.2	+4.6	+7.8
	W1	0 to 15	9.0	−0.50	−3.9	−5.5	−2.3
	W2, W5	0 to 15	9.0	−0.70	−5.4	−7.0	−3.8
Roof		0 to 15	9.0	−0.90	−6.9	−8.5	−5.3
		15 to 30	9.0	−0.50	−3.9	−5.5	−2.3
		30 to 100	9.0	−0.30	−2.3	−3.9	−0.7

Note: q_h = 9.0 psf; G = 0.85.

Table G3-71 Design Pressures for Wind Normal to Wall W2

Surface type	Surface designation	z or x (ft)	q (psf)	C_p	External pressure(psf)	Design pressures (psf) $(+GC_{pi})$	$(-GC_{pi})$
Walls	W2, W3	0 to 15	9.0	+0.80	+6.2	+4.6	+7.8
	W5	0 to 15	9.0	−0.50	−3.9	−5.5	−2.3
	W1, W4	0 to 15	9.0	−0.70	−5.4	−7.0	−3.8
Roof		0 to 15	9.0	−0.90	−6.9	−8.5	−5.3
		15 to 30	9.0	−0.50	−3.9	−5.5	−2.3
		30 to 100	9.0	−0.30	−2.3	−3.9	−0.7

Note: $q_h = 9.0$ psf; $G = 0.85$.

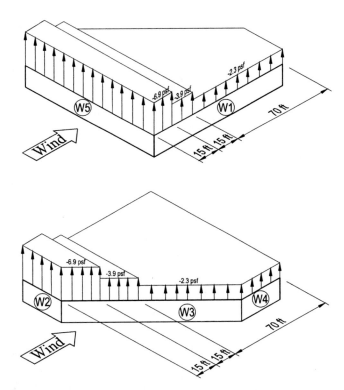

Figure G3-44 External Roof Pressure Zones for MWFRS

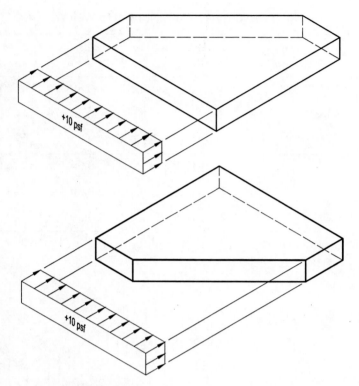

Figure G3–45   Application of 10 psf Minimum Load Case

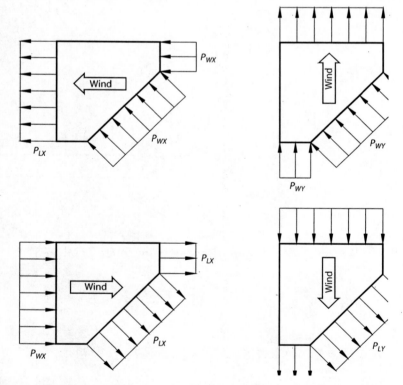

Figure G3–46   Application of Load Case 1 from Each Orthogonal Direction

126 Wind Loads: Guide to the Wind Load Provisions of ASCE 7-05

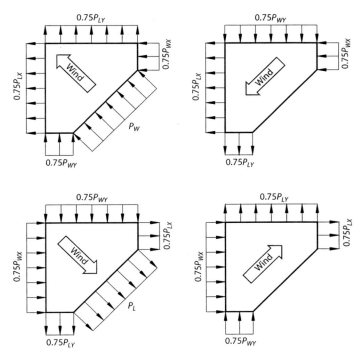

Figure G3–47   Application of Load Case 3 from Each Diagonal Direction

3.12.8   Design Pressures for C&C

Design pressure for components and cladding is obtained by

$$p = q_h[(GC_p) - (GC_{pi})] \qquad \text{(Eq. 6-22)}$$

where

q_h = 11.0 psf for Case 1

(GC_p) = External pressure coefficient (see Figures 6-11A and 6-11B of the Standard)

(GC_{pi}) = ±0.18, the internal pressure coefficient for enclosed buildings

Wall Design Pressures

The pressure coefficients (GC_p) are a function of effective wind area. Since specific components of the walls are not identified, design pressures are given for various effective wind areas in **Table G3–72**. These values have been reduced by 10% as allowed by Note 5 in Figure 6-11A for roof angle $\Theta \leq 10°$.

Width edge zone

smaller of $a = 0.1(100) = 10$ ft
or $a = 0.4(15) = 6.0$ ft (controls)
but not less than $a = 0.04(100) = 4.0$ ft
or $a = 3$ ft

Table G3–72 Wall (GC_p) for Ex.12, by Zone

A (ft^2)	Zones 4 and 5 (+GC_p)	Zone 4 (−GC_p)	Zone 5 (−GC_p)
≤10	+0.90	−0.99	−1.26
50	+0.79	−0.88	−1.04
100	+0.74	−0.83	−0.95
>500	+0.63	−0.72	−0.72

Note: GC_p values have been reduced by 10% since $\Theta \leq 10°$.

The design pressures are the algebraic sum of external and internal pressures. Controlling negative pressure is obtained with positive internal pressure, and controlling positive pressure is obtained with negative internal pressure. The controlling design pressures are given in **Table G3–73**. The edge zones for the walls are arranged at exterior corners, as shown in **Figure G3–48**.

Roof Design Pressures

The pressure coefficients (GC_p) are a function of effective wind area. Since specific components of the roof are not identified, design pressures are given for various effective wind areas in **Table G3–74**. The design pressures (**Table G3–75**) are the algebraic sum of external and internal pressures. Controlling negative pressure is obtained with positive internal pressure, and controlling positive pressure is obtained with negative internal pressure. The edge zones for the roof are arranged as shown in **Figure G3–49**.

Table G3–73 Controlling Design Pressures for Wall Components (psf), by Zone

Area	Zone 4		Zone 5	
	Positive	Negative	Positive	Negative
≤ 10	+11.9	−12.9	+11.9	−15.9
50	+10.7	−11.6	+10.7	−13.4
100	+10.1	−11.1	+10.1	−12.4
> 500	+8.9*	−9.9*	+8.9*	−9.9*

* Section 6.1.4.2 of the Standard requires that C&C pressures be not less than ± 10 psf

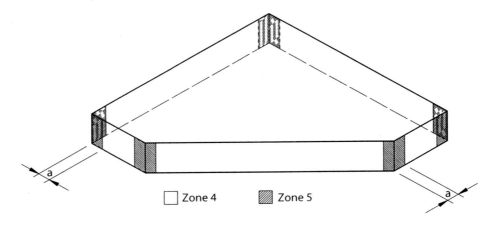

Figure G3–48 Wall Pressure Zones for Components and Cladding

Table G3–74 Roof External Pressure Coefficients, by Zone

A (ft²)	Zones 1, 2, 3 (GC_p)	Zone 1 (GC_p)	Zone 2 $(-GC_p)$	Zone 3 $(-GC_p)$
	Positive	Negative	Negative	Negative
10	+0.30	−1.00	−1.80	−2.80
50	+0.23	−0.93	−1.31	−1.61
100	+0.20	−0.90	−1.10	−1.10

Table G3–75 Roof Design Pressures (psf), by Zone

Area	Zones 1, 2, 3 (GC_p)	Zone 1 (GC_p)	Zone 2 $(-GC_p)$	Zone 3 $(-GC_p)$
	Positive	Negative	Negative	Negative
10	+5.3*	−13.0	−21.8	−32.8
50	+4.5*	−12.2	−16.4	−19.7
100	+4.2*	−11.9	−14.1	−14.1

* Section 6.1.4.2 of the Standard requires that C&C pressures be not less than ±10 psf.

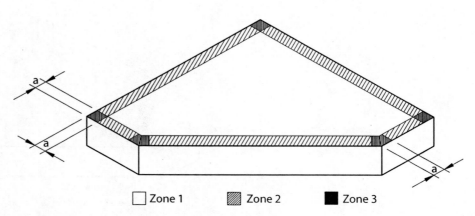

Figure G3–49 *Roof Pressure Zones for Components and Cladding*

3.13 Example 13: 30-ft × 60-ft Open Building with Gable Roof

In this example, design wind pressures for a typical, open-storage building are determined. **Figure G3–50** shows the dimensions and framing of the open buildings. The building data are as follows:

Location Tulsa, Oklahoma

Topography Homogenous

Terrain Suburban, wooded

Dimensions 30 ft × 60 ft in plan
Roof eave height is 10 ft
Roof gable $\Theta = 15°$

Framing Typical metal construction
Roof trusses spanning 30 ft are spaced 4 ft on center
Roof panels are 2 ft × 8 ft

Storage Clear flow; blockage less than 50%

3.13.1 Building Classification

The building is a minor storage building. It is not considered an essential facility and represents a low hazard to human life in the event of failure. Building Category I is appropriate; see Table 1-1 of the Standard.

3.13.2 Exposure

The building is located in a suburban area; according to Section 6.5.6.2 and 6.5.6.3 of the Standard, Exposure B is used.

3.13.3 Basic Wind Speed

Selection of basic wind speed is addressed in Figure 6-1 of Section 6.5.4 of the Standard. Therefore, the basic wind speed is $V = 90$ mph.

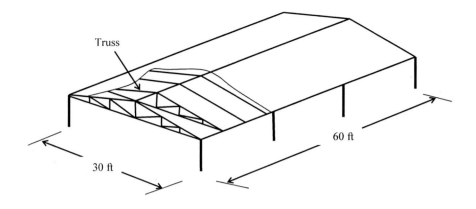

Figure G3–50 Building Characteristics for Example 13, Open Building with Gable Roof

3.13.4 Velocity Pressures

The velocity pressures are computed using

$$q_z = 0.00256 K_z K_{zt} K_d V^2 I \text{ psf} \tag{Eq. 6-15}$$

where

K_z = 0.70 from Table 6-3 of the Standard for Case 1 (C&C) and 0.57 from Table 6-3 of the Standard for Case 2; for 0 to 15 ft, there is only one value: $K_z = K_h$

K_{zt} = 1.0 for homogeneous topography (see Section 6.5.7 of the Standard)

K_d = 0.85 for buildings (see Table 6-4 of the Standard)

V = 90 mph (see Figure 6-1a of the Standard)

I = 0.87 for Category I building (see Table 6-1 of the Standard)

h = mean roof height = $10 + [(15)(\tan 15°)/2] = 12.0$ ft.

For Case 1 (C&C)

$q_h = 0.00256\ (0.7)\ (1.0)\ (0.85)\ (90)^2\ (0.87) = 10.7$ psf

For Case 2 (MWFRS)

$q_h = 0.00256\ (0.57)\ (1.0)\ (0.85)\ (90)^2\ (0.87) = 8.8$ psf

3.13.5 Design Wind Pressure for MWFRS

The equation for open buildings is given in Section 6.5.13.2 of the Standard.

$$p = q_h G C_N \tag{Eq. 6-25}$$

where

q_h = Velocity pressure evaluated at mean roof height h
G = Gust effect factor
C_N = Net pressure coefficient value obtained from Figure 6-18B and Figure 6-18D.

Gust Effect Factor

The gust effect factor for nonflexible (rigid) buildings is given in Section 6.5.8 of the Standard as $G = 0.85$.

Roof Net Pressure Coefficients

The roof net pressure coefficients (C_N) are presented in **Table G3–76**. Because the building is open and has a pitched roof, there are two wind directions to be considered: wind direction parallel to the slope (normal to ridge), $\gamma = 0°$ or 180°, and wind normal to the roof slope (parallel to ridge), $\gamma = 90°$. For $\gamma = 0°$ or 180° the net pressure coefficients are obtained from Figure 6-18B. For $\gamma = 90°$ the net pressure coefficients are obtained from Figure 6-18D.

Clear wind flow is assumed; blockage less than 50%.

Wind Pressure for MWFRS

Calculated wind pressures for MWFRS are summarized in **Table G3–77**.

$$p = q_h G C_N = 8.8 \times 0.85 \times C_N$$

Figures G3–51 and **G3–52** illustrate the design pressures for Case A and Case B for wind directions $\gamma = 0°$ or 180° and $\gamma = 90°$, respectively.

Minimum Design Wind Loadings

Section 6.1.4.1 of the Standard requires that the design wind load for MWFRS of open buildings shall not be less than 10 psf (net horizontal) multiplied by area projected on a plane normal to the wind direction. Depending on the

Table G3–76 Roof Net Pressure Coefficients (C_N) for Two Cases

Wind direction	$\theta°$	Distance from Windward Edge	Case A	Case B
Normal to ridge $\gamma = 0°$ or 180°; Figure 6–18B)	15		1.1 (C_{NW}) −0.4 (C_{NL})	0.1 (C_{NW}) −1.1 (C_{NL})
Parallel to ridge ($\gamma = 90°$; Figure 6–18D)	0	0–12 ft	−0.8	0.8
		12–24 ft	−0.6	0.5
		24–60 ft	−0.3	0.3

Table G3–77   Design wind pressures for MWFRS (psf)

Wind Direction, γ	Distance from Windward Edge	Case A	Case B
Normal to ridge ($\gamma = 0°$ or $180°$) Windward		8.2	0.7
Leeward		–3.0	–8.2
Parallel to ridge ($\gamma = 90°$)	0–12 ft	–6.0	6.0
	12–24 ft	–4.5	3.7
	24–60 ft	–2.2	2.2

Note: Positive number means toward the surface; negative numbers mean away from the surface.

projected area of the roof and supporting structure, this minimum loading could govern and should be checked. Load Case A satisfies this minimum requirement.

3.13.6 Design Pressures for Roof Trusses and Roof Panels

The design pressure equation for C&C for open buildings with pitched roofs is given in Section 6.5.13.3 of the Standard.

$$p = q_h G C_N \qquad \text{(Eq. 6-26)}$$

where

q_h = Velocity pressure evaluated at mean roof height h

G = Gust effect factor value determined from Section 6.5.8 of the Standard

C_N = Net pressure coefficient value obtained from Figure 6-19B.

Roof Design Pressures

Roof net design coefficients for components and cladding are presented in **Table G3–78**, from Figure 6-19B in the Standard.

Width of Zone 2 and Zone 3

    Smaller of $a = 0.4(12) = 4.8$ ft
    or $a = 0.1(30) = 3$ ft (controls)
    but not less than $a = 0.04(30) = 1.2$ ft
    or $a = 3$ ft
        $a^2 = 9$ sq ft

Effective wind area

    Roof panel $A = 4 \times 2 = 8$ sq ft (controls)
    or $A = 4 \times (4/3) = 5.3$ sq ft
    Roof truss $A = 4 \times 30 = 120$ sq ft
    or $A = 30 \times (30/3) = 300$ sq ft (controls)

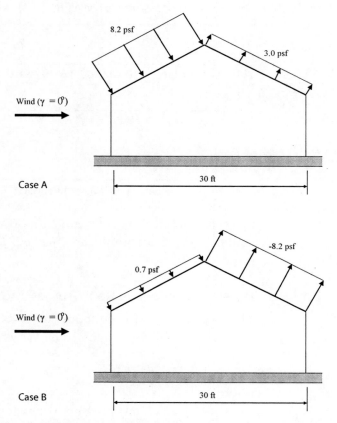

Figure G3–51   Design Pressures for MWFRS for Wind Direction $\gamma = 0°$ or $180°$. Note: a) Case A and b) Case B.

Wind Pressure for Trusses and Roof Panels

Calculated wind pressures for trusses and roof panels are summarized in **Table G3–79**.

$$p = q_h G C_N = 10.7 \times 0.85 \times C_N$$

Zones for the pitched roof of this open building are shown in **Figure G3–53**. The panels and trusses are designed for the pressures indicated.

For trusses, two loading combinations need to be considered. The two loading cases are shown in **Figure G3–54**. Loading case 1 is for wind directions 1 and 2 while loading case 2 is for wind direction 2. The loadings shown for trusses are used for the design of truss and individual members. For anchorage of the truss to frame support members may use MWFRS loading.

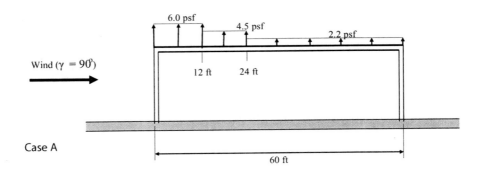

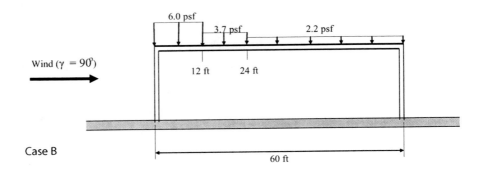

Figure G3-52 Design Pressures for MWFRS for Wind Direction γ= 90°. Note: a) Case A and b) Case B. The pressures are normal to roof surface

Table G3-78 Roof Net Pressure Coefficients (CN) for Components, θ = 15°, by Zone

Component	Effective Wind Area (sq ft)	Zone 3		Zone 2		Zone 1	
Panel	8 ($\leq a^2$)	2.2	−2.2	1.7	−1.7	1.1	−1.1
Truss	300 ($> 4a^2$)	1.1	−1.1	1.1	−1.1	1.1	−1.1

Note: Coefficients are from Figure 6–19B of the Standard.

Table G3-79 Roof Component Design Pressure (psf), by Zone

Component	Zone 3		Zone 2		Zone 1	
Panel	20	−20	15.5	−15.5	10	−10
Truss	10	−10	10	−10	10	−10

Note: Positive number means toward the surface; negative numbers mean away from the surface.

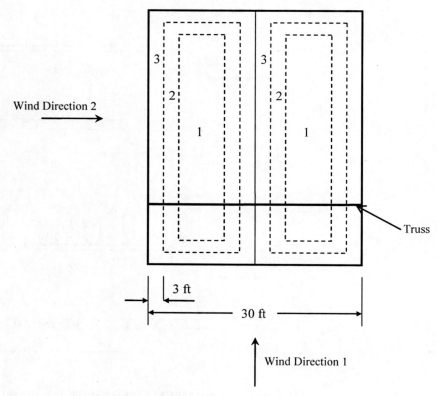

Figure G3–53 Pressure Zones for Panels and Trusses

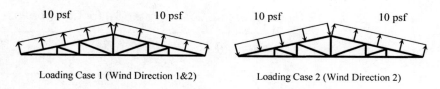

Figure G3–54 Loading Cases for Roof Trusses

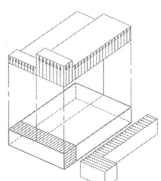

Chapter 4
Frequently Asked Questions

Over the last several years, the authors have fielded hundreds of questions and inquiries from users of the ASCE 7 wind load provisions. The purpose of this chapter is to clarify provisions of the Standard about which questions frequently and repeatedly arise. The questions are assembled by subject matter of wind load calculations to make it easier to find them.

4.1 Wind Speed

1. Is it possible to obtain larger scale maps of basic wind speeds (see Figures 6-1, 6-1A, 6-1B, and 6-1C) so that the locations of the wind speed contours can be determined with greater accuracy?

    No. The wind speed contours in the hurricane-prone region of the United States are based on hurricane wind speeds from Monte Carlo simulations and on estimates of the rate at which hurricane wind speeds attenuate to 90 mph following landfall. Because the wind speed contours of these figures represent a consensus of the ASCE 7 Wind Load Task Committee, increasing the map scale would do nothing to improve their accuracy. Some jurisdictions establish a specific basic wind speed for a jurisdictional area; it is advisable to check with authorities of that area.

2. IBC (2006) Figure 1609 gives the 3-s wind speed at the project location. However, according to the Notes, Figure 1609 is for Exposure C. If the project location is Exposure B, what is the proper wind speed to use?

Basic wind speed in IBC Figure 1609 or ASCE 7-05 is defined as 3-s gust wind speed at 33 ft above ground for Exposure Category C, which is the standard measurement. The velocity pressure exposure coefficient, K_z, adjusts the wind speed for exposure and height above ground. However, for simplicity the coefficient is applied in the pressure equation, thus adjusting pressure rather than wind speed. Use of K_z adjusts the pressures from Exposure C to Exposure B.

3. If the design wind loads are to be determined for a building that is located in a special wind region (shaded areas) in Figures 6-1, 6-1A, 6-1B, and 6-1C, what basic wind speed should be used?

The purpose of the special wind regions in these figures is to alert the designer to the fact that there are regions in which wind speed anomalies are known to exist. Wind speeds in these regions may be substantially higher than the speeds indicated on the map, and the use of regional climatic data and consultations with a wind engineer or meteorologist are advised.

4. How do I design for a Category 3 hurricane?

The Saffir/Simpson Hurricane Scale classifies hurricanes based on intensity and damage potential using five categories (1 through 5, with 5 being the most intense). Table C6-2 in the Commentary of the standard gives an approximate relationship between wind speed in ASCE 7 and the Saffir-Simpson hurricane scale. The Saffir-Simpson scale has a range of wind speeds. A decision will have to be made on which specific basic wind speed over land to be used.

4.2 Load Factor

5. When can I use the one-third stress increase specified in some material standards?

When using the loads or load combinations specified in ASCE 7-05, no increase in allowable stress is permitted except when the increase is justified by the rate of duration of load (such as duration factors used in wood design). Instead, load combination #6 in Section 2.4.1 of ASCE 7-05 adjusts for the case when wind load and another transient load are combined. This load combination applies a 0.75 factor to the transient loads ONLY (not to the dead load). The 0.75 factor applied to the transient loads accounts for the fact that it is extremely unlikely that two maximum events will happen at the same time.

6. **Why can the wind directionality factor (K_d) only be used with the load combinations specified in Sections 2.3 and 2.4 of ASCE 7-05?**

   In the strength design load combinations provided in previous editions of ASCE 7 (ASCE 7-95 and earlier), the 1.3 factor for wind included a "wind directionality factor" of 0.85. In ASCE 7-98, the loading combinations used 1.6 instead of 1.3 (approximately equals 1.6×0.85), and the directionality factor is included in the equation for velocity pressure. Separating the directionality factor from the load combinations allows the designer to use specific directionality factors for each structure and allows the factor to be revised more readily when new research becomes available.

4.3 Terrain Exposure

7. **What exposure category should I use for the main wind force-resisting systems (MWFRS) if the terrain around my site is Exposure B, but there is a large parking lot directly next to one of the elevations?**

   Section 6.5.6 of ASCE 7-05 provides general definitions of Exposures B, C, and D; however, the designer must refer to the Commentary for a detailed explanation for each exposure. The exposure depends on the size of the parking lot, its size relative to the building, and the number and type of obstructions in the area. Section C6.5.6 of the Commentary includes a formula (Eq. C6-3) that will help the designer determine if the terrain roughness is sufficient to be categorized as Exposure B. Note that, for Exposure B, the fetch distance is 2,600 ft or 20 times the structure's height, whichever is greater. Also note that the Commentary provides suggestions for determining the "upwind fetch surface area."

   For clearings such as parking lots, wide roads, road intersections, underdeveloped lots, and tree clearings, the Commentary provides a rational procedure and an example to interpolate between Exposure B and C; the designer is encouraged to use this procedure. The procedure is illustrated in Figures C6-8 and C6-9. The determination of exposure category is based on the extent and the distance from the project location of "open patches" where the exposure could be considered to be less rough than Exposure B.

8. **Under what conditions is it necessary to consider speed-up due to topographic effects when calculating wind loads?**

   Section 6.5.7 of the Standard requires the calculation of the topographic factor, K_{zt}, for buildings and other structures sited on the upper half of isolated hills or escarpments located in Exposures B, C, or D where the upwind terrain is free of such topographic features for a distance of at least 100 H or 2 mi, whichever is smaller, as measured from the crest of the topographic feature. K_{zt} need not be calculated when the height, H, is less than 15 ft in Exposures D and C, or less than 60 ft in Exposure B. In addition, K_{zt} need not be calculated when H/L_h is less than 0.2. H and L_h are defined in Figure 6-4. The value of K_{zt} is never less than 1.0.

4.4 MWFRS and C&C

9. **Do I consider a tilt-up wall system to be components and cladding (C&C) or main wind force-resisting systems (MWFRS) or both?**

    Both. Depending on the direction of the wind, a tilt-up wall system must resist either MWFRS forces or C&C forces. In the C&C scenario, the elements receive the wind pressure directly and transfer the forces to the MWFRS in the other direction. When a tilt-up wall acts as a shear wall, it is resisting forces of MWFRS. Because the wind is not expected to blow from both directions at the same time, the MWFRS forces and C&C forces are analyzed independently from each other in two different load cases. This is also true of masonry and reinforced-concrete walls.

10. **When is a gable truss in a house part of the MWFRS? Should it also be designed as a C&C? What about individual members of a truss?**

    Roof trusses are considered to be components since they receive load directly from the cladding. However, since a gable truss receives wind loads from more than one surface, which is part of the definition for MWFRS, an argument can be made that the total load on the truss is more accurately defined by the MWFRS loads. A common approach is to design the members and internal connections of the gable truss for C&C loads, while using the MWFRS loads for the anchorage and reactions of the truss with wall or frame. When designing shear walls or cross-bracing, roof loads can be considered as MWFRS.

    In the case where the tributary area on any member exceeds 700 ft^2, Section 6.5.12.1.3 permits the member to be considered an MWFRS for determination of wind load. However, for a gable truss as MWFRS under this provision, the top chord members of the truss would have to follow rules of C&C if they receive load directly from the roof sheathing.

4.5 Gust Effect Factor

11. A tower has a fundamental frequency of 2 Hz, but has a height-to-width ratio of 6. Should the tower be treated as a flexible structure to determine the gust effect factor?

No. Definition of a flexible building given in Section 6.2 states that fundamental (first mode) natural frequency less than 1 Hz would make it flexible for the gust effect factor. The energy in the turbulence spectrum is small for frequencies above 1 Hz. Hence, a tower with fundamental frequency of 2 Hz is not likely to be dynamically excited by wind. The commentary of the Standard has a good discussion on response of buildings and structures in turbulent wind in Section C6.5.8.

4.6 Pressure Coefficient

12. In the design of main wind force-resisting systems (MWFRS), the provisions of Figure 6-6 apply to enclosed or partially enclosed buildings of all heights. The provisions of Figure 6-10 apply to enclosed or partially enclosed buildings with mean roof height less than or equal to 60 ft. Does this mean that either figure may be used for the design of a low-rise MWFRS?

Figure 6-6 may be used for buildings of any height, whereas Figure 6-10 applies only to low-rise buildings. Section 6.2 defines low-rise buildings to comply with mean roof height $h \leq 60$ ft and h not to exceed least horizontal dimensions. Pressure coefficients for low-rise buildings given in Figure 6-10 represent "pseudo" loading conditions enveloping internal structural reactions of total uplift, total horizontal shear, bending moment, etc. (see Section C6.5.11). Thus, they are not real wind-induced loads. These loads work adequately for buildings of the shapes shown in Figure 6-10; they are not applicable to other shapes such as the U-shaped building in Example 9 or the odd-shaped building of Example 12.

13. What pressure coefficients should be used to reflect contributions for the underside (bottom) of the roof overhangs and balconies?

Sections 6.5.11.4.1 and 6.5.11.4.2 specify pressure coefficients to be used for roof overhangs to determine loads for MWFRS and C&C, respectively. No specific guidance is given for balconies, but use of the loading criteria for roof overhangs should be adequate.

Other loading conditions at the overhang, such as the pressures on the upper roof surface, must be considered to obtain the total loads for connections between the roof and the wall.

14. **If the mean roof height, h, is greater than 60 ft with a roof geometry that is other than flat roof, what pressure coefficients are to be used for roof C&C design loads?**

    Section 6.5.12.4.3 permits use of pressure coefficients of Figures 6-11 through 6-16 provided the mean roof height $h < 90$ ft, the height-to-width ratio is 1 or less, and Eq. 6-22 is used.

    Note 6 of Figure 6-17 permits use of coefficients of Figure 6-11 when the roof angle $\theta > 10°$.

15. **Flat roof trusses are 30 ft long and are spaced on 4-ft centers. What effective wind area should be used to determine the design pressures for the trusses?**

    Roof trusses are classified as C&C since they receive wind load directly from the cladding (roof sheathing). In this case, the effective wind area is the span length multiplied by an effective width that need not be less than one-third the span length or $(30)(30/3) = 300$ ft². This is the area on which the selection of GC_p should be based. Note, however, that the resulting wind pressure acts on the tributary area of each truss, which is $(30)(4) = 120$ ft².

16. **Roof trusses have a clear span of 70 ft and are spaced 8 ft on center. What effective wind area should be used to determine the design pressures for the trusses?**

    According to the definition of Effective Wind Area in Section 6.2 the effective wind area is $(70)(70/3) = 1,633$ ft². The tributary area of the truss is $(70)(8) = 560$ ft², which is less than the 700-ft² area required by Section 6.5.12.1.3 to qualify for design of the truss using the rules for MWFRS. The truss is to be designed using the rules for C&C, and the wind pressure corresponding to an effective wind area of 1,633 ft² is to be applied to the tributary area of 560 ft².

17. **Metal decking consisting of panels 20 ft long and 2 ft wide is supported on purlins spaced 5 ft apart. Will the effective wind area be 40 ft² for the determination of pressure coefficients?**

    Although the length of a decking panel is 20 ft, the basic span is 5 ft. According to the definition of effective wind area, this area is the span length multiplied by an effective width that need not be less than one-third the span length. This gives a minimum effective wind area of $(5)(5/3) = 8.3$ ft². However, the actual width of a panel is 2 ft, making the effective wind area equal to the tributary area of a single panel, or $(5)(2) = 10$ ft². Therefore, GC_p would be determined on the basis of 10 ft² of effective wind area, and the corresponding wind load would be applied to a tributary area of 10 ft². Note that GC_p is constant for effective wind areas less than 10 ft².

18. A masonry wall is 12 ft in height and 80 ft long. It is supported at the top and at the bottom. What effective wind area should be used in determining the design pressure for the wall?

    For a given application, the magnitude of the pressure coefficient, GC_p, increases with decreasing effective wind area. Therefore, a very conservative approach would be to consider an effective wind area with a span of 12 ft and a width of 1 ft, and design the wall element as C&C. However, the definition of effective wind area states that this area is the span length multiplied by an effective width that need not be less than one-third the span length. Accordingly, the effective wind area would be $(12)(12/3) = 48$ ft².

19. If the pressure or force coefficients for various roof shapes (e.g., a canopy) are not given in ASCE 7-05, how can the appropriate wind forces be determined for these shapes?

    With the exception of pressure or force coefficients for certain shapes, parameters such as V, I, K_z, K_{zt}, and G are given in ASCE 7-05. It is possible to use pressure or force coefficients from the published literature (see Section 1.4 of this guide) provided these coefficients are used with care. Mean pressure or force coefficients from other sources can be used to determine wind loads for MWFRS. However, it should be recognized that these coefficients might have been obtained in wind tunnels that have smooth, uniform flows as opposed to more proper turbulent boundary-layer flows. Pressure coefficients for components and cladding obtained from the literature should be adjusted to the 3-s gust speed, which is the basic wind speed adopted by ASCE 7-05.

20. Can the pressure/force coefficients given in ASCE 7-05 be used with the provisions of ASCE 7-88, 7-93, 7-95, 7-98, or 7-02?

    No. ASCE 7-88 (and 7-93) used the fastest-mile wind speed as the basic wind speed. With the adoption of the 3-s gust speed starting with ASCE 7-95, the values of certain parameters used in the determination of wind forces have been changed accordingly. The provisions of ASCE 7-88 and 7-05 should not be interchanged. Coefficients in ASCE 7-95, 7-98, 7-02, and 7-05 are consistent; they are related to 3-s gust speed.

21. What is the difference between a partially enclosed and an enclosed building, and how are they determined?

    The definition of these two enclosure categories is in Section 6.2. There is a specific definition for both partially enclosed and open buildings; an enclosed building is one that does not comply with the requirements for either open or partially enclosed; "enclosed" is a default designation. The determination of a partially enclosed condition is a function of the area of a windward wall dominant opening compared to the area of openings in the remainder of the building envelope.

It is possible to have a building be classified as an enclosed building even when there are large openings in two or more walls and when it does not fit the definition of a partially enclosed building.

The significance in the determination of wind pressure is that partially enclosed buildings have an internal pressure coefficient $GC_{pi} = \pm 0.55$ and enclosed buildings have a $GC_{pi} = \pm 0.18$.

4.7 Force Coefficient

22. **What constitutes an open building? If a process plant has a three-story frame with no walls but with a lot of equipment inside the framing, is this an open building?**

    An open building is a structure in which each wall is at least 80% open (see Section 6.2). Yes, this three-story frame would be classified as an open building, or as "other" structure. In calculating the wind force, F, appropriate values of C_f and A_f would have to be assigned to the frame and to the equipment inside.

23. **A trussed tower of 10- × 10-ft² cross section consists of structural angles forming basic tower panels 10 ft high. The solid area of the face of one tower panel projected on a plane of that face is 22 ft². What force coefficient, C_f, should be used to calculate the wind force? What would the force coefficient be for the same tower fabricated of rounded members and having the same projected solid area? What area should be used to obtain the wind force per foot of tower height acting (1) normal to a tower face, and (2) along a tower diagonal?**

    The gross area of one panel face is $(10)(10) = 100$ ft², and the solidity ratio is $\epsilon = 22/100 = 0.22$. For a tower of square cross section, the force coefficient from Figure 6-23 is as follows:

    $$C_f = (4)(0.22)^2 - (5.9)(0.22) + 4.0 = 2.90$$

    For rounded members, the force coefficient may be reduced by the factor

    $$(0.51)(0.22)^2 + 0.57 = 0.59$$

    Thus, the force coefficient for the same tower constructed of rounded members with the same projected area would be

$C_f = (0.59)(2.90) = 1.71$

The area, A_f, used to calculate the wind force per foot of tower height is $22/10 = 2.2$ ft^2 for all wind directions.

24. In calculating the wind forces acting on a trussed tower of square cross section (see Figure 6-23), should the force coefficient, C_f, be applied to both the front and the back (windward and leeward) faces of the tower?

    No. The calculated wind forces are the total forces acting on the tower. The force coefficients given in Figure 6-23 include the contributions of both front and back faces of the tower, as well as the shielding effect of the front face on the back face.

4.8 Miscellaneous

25. Section 6.1.4.1 provides for a minimum wind pressure of 10 lb/ft^2 multiplied by the area of the building or structure projected onto a vertical plane normal to the assumed wind direction of MWFRS. Does this provision apply to low-rise buildings?

    Yes, it should. There was some confusion in ASCE 7-98 provisions for low-rise buildings where it was difficult to interpret application of loads on building frames using the two cases of loads at each corner. In ASCE 7-05, application of loads on low-rise buildings is clarified with illustrative sketches, and only one table of pressure coefficients is provided (See Figure 6-10 of the standard). In addition, Note 6 is added to account for minimum total horizontal shear, although this provision does not guarantee minimum 10 psf on the projected area of the building.

26. Equation 6-15 for velocity pressure uses the subscript *z* while other equations use subscripts *z* and *h*. When is *z* used and when is *h* used?

    Equation 6-15 is the general formula for the velocity pressure, q_z, at any height, *z*, above ground. There are many situations in the Standard where a specific value of *z* is called for, namely the height (or mean roof height) of a building or other structure. Whenever the subscript *h* is called for, it is understood that *z* becomes *h* in the appropriate equations.

27. Section 6.2 of the Standard provides definitions of glazing, impact resistant; impact-resistant covering; and wind-borne debris regions. To be impact resistant, the Standard specifies that the glazing of the building envelope must be shown by an approved test method to withstand the impact of wind-borne missiles likely

to be generated during design winds. Where does one find information on appropriate test methods?

Section 6.5.9.3 of the standard refers to two standards, ASTM E1886 and ASTM E1996. References for these standards are given in Section 6.7 of the 7-05 standard. The two standards specify test methods and performance standards.

28. **The Standard does not provide for across-wind excitation caused by vortex shedding. How can one determine when vortex shedding might become a problem?**

Vortex shedding is almost always present with bluff-shaped cylindrical bodies. Vortex shedding can become a problem when the frequency of shedding is close to or equal to the frequency of the first or second transverse frequency of the structure. The intensity of excitation increases with aspect ratio (height-to-width or length-to-breadth) and decreases with increasing structural damping. Structures with low damping and with aspect ratios of eight or more may be prone to damaging vortex excitation. If across-wind or torsional excitation appears to be a possibility, expert advice should be obtained.

29. **If high winds are accompanied by rain, will the presence of raindrops increase the mean density of the air to the point where the wind loads are affected?**

No. Although raindrops will increase the mean density of the air, the increase is small and may be neglected. For example, if the average rate of rainfall is 5 in./h, the average density of raindrops will increase the mean air density by less than 1%.

30. **What wind loads do I use during construction?**

ASCE 7 does not address wind loads during construction. Construction loads are specifically addressed in the standard ASCE/SEI 37-02, *Design Loads on Structures during Construction* (ASCE 2002).

31. **Is it possible to determine the wind loads for the design of interior walls?**

The Standard does not address the wind loads to be used in the design of interior walls or partitions. A conservative approach would be to apply the internal pressure coefficients $GC_{pi} = \pm 0.18$ for enclosed buildings and $GC_{pi} = \pm 0.55$ for partially enclosed buildings. Post-disaster surveys have revealed the failure of interior walls when the building envelope has been breached.

References

Akins, R.E., and Cermak, J.E. (1975). *Wind pressures on buildings,* Technical Report CER7677REA-JEC15, Fluid Dynamics and Diffusion Lab., Colorado State Univ., Fort Collins, Colo.

ANSI (1972). *Minimum design loads for buildings and other structures,* ANSI A58.1-1972, ANSI, New York.

ANSI (1982). *Minimum design loads for buildings and other structures,* ANSI A58.1-1982, ANSI, New York.

ANSI/EIA/TIA (1991). *Structural standards for steel antenna towers and antenna supporting structures,* ANSI/EIA/TIA-222-E, Electronic Industries Assn., Washington, D.C.

ASCE (1961). "Wind forces on structures." *Transactions,* 126(2), 1124–1198.

ASCE (1990). *Minimum design loads for buildings and other structures,* ASCE/ANSI 7-88, ASCE, New York.

ASCE (1996). *Minimum design loads for buildings and other structures,* ASCE/ANSI 7-95, ASCE, New York.

ASCE (1997). *Wind loads and anchor bolt design for petrochemical facilities,* Report from the Committees on Wind Induced Forces and on Anchor Bolt Design, ASCE, New York.

ASCE (1999). *Wind tunnel studies of buildings and structures,* N. Isyumov, ed. Manual of Practice No. 67, ASCE, Reston, Va.

ASCE (2000). *Minimum design loads for buildings and other structures,* ASCE 7-98, ASCE, Reston, Va.

ASCE (2002). *Design loads on structures during construction,* ASCE-SEI 37-02, ASCE, Reston, VA.

ASCE (2006). *Minimum design loads for buildings and other structures,* ASCE 7-05, ASCE, Reston, Va.

ASTM (2005). *Standard test method for performance of exterior windows, curtain walls, doors, and impact protective systems impacted by missile(s) and exposed to cyclic pressure differentials,* ASTM E1886-05, ASTM, West Conshohocken, Pa.

ASTM (2009). *Standard specification for performance of exterior windows, curtain walls, doors, and impact protective systems impacted by windborne debris in hurricanes,* ASTM E1996-09, ASTM, West Conshohoken, Pa.

Batts, M.E., Cordes, M.R., Russell, L.R., Shaver, J.R., and Simiu, E. (1980). *Hurricane wind speeds in the United States,* NBS Building Science Series 124, National Bureau of Standards, Washington, DC.

Behr, R.A., and Minor, J.E. (1994). "A survey of glazing system behavior in multi-story buildings during Hurricane Andrew." *The Structural Design of Tall Buildings,* 3, 143–161.

Best, R.J., and Holmes, J.D. (1978). *Model study of wind pressures on an isolated single-story house,* Wind Engrg. Report 3/78, James Cook Univ. of North Queensland, Australia.

British Standards Institute (1997). *Loading for buildings, part 2: Code of practice for wind loads,* BS 6399-2, British Standards Institute, London.

Cermak, J.E. (1977). "Wind-tunnel testing of structures." *J. Engrg. Mech. Div.,* 103(6), 1125–1140.

Cook, N.J. (1985). *The designer's guide to wind loading of building structures,* Parts I and II. Butterworth Publishers, London.

Davenport, A.G., Surry, D., and Stathopoulos, T. (1977). *Wind loads on low-rise buildings,* Final report on phases I and II, BLWT-SS8, Univ. of Western Ontario.

Davenport, A.G., Surry, D., and Stathopoulos, T. (1978). *Wind loads on low-rise buildings,* Final report on phase III, BLWT-SS4, Univ. of Western Ontario.

Durst, C.S. (1960). "Wind speeds over short periods of time." *Meteorological Magazine,* 89, 181–187.

Eaton, K.J., and Mayne, J.R. (1975). "The measurement of wind pressures on two-story houses at Aylesbury." *J. Industrial Aerodynamics,* 1(1), 67–109.

Eurocode 1 (1998). *Basis of design and actions on structures/wind action, part 2.4: Wind actions,* CEN/TC 250/SC1, Technical Secretariat, Brussels.

FEMA (1980). *Interim guidelines for building occupant protection from tornadoes and extreme winds,* FEMA TR83-A, FEMA, Washington, D.C.

FEMA (2008a). *Design and construction guidance for community safe rooms,* Publication 361, FEMA, Washington, D.C.

FEMA (2008b). *Taking shelter from the storm: Building a saferoom inside your home,* Publication 320, FEMA, Washington, D.C.

Georgiou, P.N., Davenport, A.G., and Vickery, B.J. (1983). "Design wind speeds in regions dominated by tropical cyclones." *J. Wind Engrg. and Industrial Aerodynamics,* 13, 139–152.

Ho, E. (1992). "Variability of low building wind loads," Ph.D. thesis, Univ. of Western Ontario.

Hoerner, S.F. (1965). *Fluid dynamics drag,* Pub. by S.F. Hoerner, Midland Park, N.J.

Holmes, J.D. (2001). *Wind loading of structures,* Spon Press, New York.

Holmes, J.D., Melbourne, W.H., and Walker, G.R. (1990). "A commentary on the Australian standard for wind loads," Australian Wind Engrg. Society.

ICC (2006), *International building code,* International Code Council, Country Club Hills, Ill.

ISO (1997). *Wind actions on structures,* ISO 4354, International Standards Org., Geneva.

Isyumov, N. (1982). "The aeroelastic modeling of tall buildings." *Proceedings, international workshop on wind tunnel modeling criteria and techniques in civil engrg. applications,* NBS, Gaithersburg, Md., 373–407.

Isyumov, N., and Case, P. (1995). *Evaluation of structural wind loads for low-rise buildings contained in ASCE Standard 7-95,* BLWT-SS17-1995, Univ. of Western Ontario.

Jackson, P.S., and Hunt, J.C.R. (1975). "Turbulent wind flow over a low hill." *Qtrly. J. Royal Meteorological Society,* 101, 929–955.

Kareem, A. (1985). "Lateral-torsional motion of tall buildings." *J. Struct. Engrg.,* 111(11), 2479–2496.

Kareem, A. (1992). "Dynamic response of high-rise buildings to stochastic wind loads." *J. Wind Engrg. and Industrial Aerodynamics,* 41–44.

Kareem, A., and Smith, C.E. (1994). "Performance of offshore platforms in Hurricane Andrew." *Hurricanes of 1992: Lessons learned and implications for the future,* R.A. Cook and M. Sotani, eds., ASCE, New York.

Kavanagh, K.T., Surry, D., Stathopoulos, T., and Davenport, A.G. (1983). *Wind loads on low-rise buildings: Phase IV,* BLWT-SS14, Univ. of Western Ontario.

Krayer, W.R., and Marshall, R.D. (1992). "Gust factors applied to hurricane winds." *Bull. American Meteorological Society,* 73, 613–617.

Lawson, T.V. (1980). *Wind effects on buildings* (Vols. 1 and 2), Applied Science Publishers Ltd., Essex, England.

Lemelin, D.R., Surry, D., and Davenport, A.G. (1988). "Simple approximations for wind speed-up over hills." *J. Wind Engrg. and Industrial Aerodynamics,* 28, 117–127.

Liu, Henry (1991). *Wind engineering: A handbook for structural engineers,* Prentice-Hall, New York.

Marshall, R.D., and Yokel, F.Y. (1995). *Recommended performance-based criteria for the design of manufactured home foundation systems to resist wind and seismic loads,* NISTIR 5664, NIST, Gaithersburg, Md.

McDonald, J.R. (1983). *A methodology for tornado hazard probability assessment,* NUREG/CR3058, U.S. Nuclear Regulatory Comm., Washington, D.C.

Mehta, K.C. and Delahay, J.M. (2004). *Guide to the use of the wind load provisions of ASCE 7-02*, ASCE, Reston, Va.

Mehta, K.C. and Marshall, R.D. (1998). *Guide to the use of the wind load provisions of ASCE 7-95*, ASCE, Reston, Va.

Mehta, K.C. and Perry, D.C (2001). *Guide to the use of the wind load provisions of ASCE 7-98*, ASCE, Reston, Va.

Miami/Dade County Building Code Compliance Office. "Criteria for testing products subject to cyclic wind pressure loading." *Protocol PA 203-94*.

Miami/Dade County Building Code Compliance Office. "Impact test procedures." *Protocol PA 201-94*.

Minor, J.E. (1982). "Tornado technology and professional practice." *J. Struct. Div.*, 108(11), 2411–2422.

Minor, J.E., and Behr, R.A. (1994). "Improving the performance of architectural glazing in hurricanes." *Hurricanes of 1992: Lessons learned and implications for the future*, R.A. Cook and M. Sotani, eds., ASCE, New York.

Minor, J.E., McDonald, J.R., and Mehta, K.C. (1993). *The tornado: An engineering oriented perspective*, NOAA Technical Memorandum, National Weather Service SR-147 (Reprint of Technical Memorandum ERL NSSL-82), National Oceanic and Atmospheric Admin., Washington, D.C.

Murray, R.C., and McDonald, J.R. (1993). "Design for containment of hazardous materials." *Geophysical Monograph 79: The tornado: its structure, dynamics, prediction and hazards*, C. Church, D. Burgess, C. Doswell, and R. Davies-Jones, eds. American Geophysical Union, Washington, D.C., 379–387.

Newberry, C.W., and Eaton, K.J. (1974). *Wind loading handbook*, Building Research Establishment Report K4F. Her Majesty's Stationery Office, London.

NRCC (2005). *National building code of Canada 2005*. Assoc. Committee on the National Building Code of Canada, National Research Council Canada, Ottawa, Ont.

Peterka, J.A. (1992). "Improved extreme wind prediction for the United States." *J. Wind Engrg. and Industrial Aerodynamics*, 41, 533–541.

Peterka, J.A., and Cermak, J.E. (1975). "Wind pressures on buildings: Probability densities." *J. Struct. Div.*, 101(6), 1255–1267.

Peterka, J.A., and Shahid, S. (1993). "Extreme gust wind speeds in the U.S." *Proceedings, 7th U.S. national conference on wind engineering*, UCLA, Los Angeles, 503–512.

Reinhold, T.A., ed. (1982). "Wind tunnel modeling for civil engineering applications." *Proceedings, international workshop on wind tunnel modeling criteria and techniques in civil engineering applications*, NBS, Gaithersburg, Md.

SAA (1989). *Australian standard SAA loading code, part 2: wind loads.* Standards Australia, Sydney.

Saathoff, P.J., and Stathopoulos, T. (1992). "Wind loads on buildings with sawtooth roofs." *J. Struct. Engrg.*, 118(2), 429–446.

SBCCI (1999). *SBCCI test standard for determining impact resistance from windborne debris,* SSTD 12-99, Southern Building Code Congress International, Birmingham, Ala.

Simiu, E. (1981). "Modern developments in wind engineering: Parts 1–4." *Engrg. Structures*, 3.

Simiu, E., and Scanlan, R.H. (1996). *Wind effects on structures,* 3rd ed., Wiley, New York.

Solari, G. (1993a). "Gust buffeting I: Peak wind velocity and equivalent pressure." *J. Struct. Engrg.*, 119(2), 365–382.

Solari, G. (1993). "Gust buffeting II: Dynamic along-wind response." *J. Struct. Engrg.*, 119(2), 383–398.

Standards Australia (2002). *Structural design actions: Wind actions,* AS/NZS 1170.2, Standards Australia, Sydney.

Stathopoulos, T. (1981). "Wind loads on eaves of low buildings." *J. Struct. Engrg. Div.*, 107(10), 1921–1934.

Stathopoulos, T., and Dumitrescu-Brulotte, M. (1989). "Design recommendations for wind loading on buildings of intermediate height." *Canadian J. Civil Engrg.*, 16(6), 910–916.

Stathopoulos, T., and Luchian, H.D. (1990). "Wind pressures on building configurations with stepped roofs." *Canadian J. Civil Engrg.*, 17(4), 569–577.

Stathopoulos, T., and Mohammadian, A.R. (1986). "Wind loads on low buildings with mono-sloped roofs." *J. Wind Engrg. and Industrial Aerodynamics*, 23, 81–97.

Stathopoulos, T., and Saathoff, P. (1991). "Wind pressures on roofs of various geometries." *J. Wind Engrg. and Industrial Aerodynamics*, 38, 273–284.

Surry, D., and Stathopoulos, T. (1988). *The wind loading of buildings with monoslope roofs,* Final report, BLWT-SS38, Univ. of Western Ontario.

TDI (1998). "Test for impact and cyclic wind pressure resistance of impact protective systems and exterior opening systems," Appendix E. *Building Code for Windstorm Resistant Construction, TDI Standard 1-98,* Texas Dept. of Insurance, Austin.

Vickery, P.J., and Twisdale, L.S. (1993). "Prediction of hurricane wind speeds in the U.S." *Proceedings, 7th U.S. national conference on wind engineering*, UCLA, Los Angeles, 823–832.

Walmsley, J.L., Taylor, P.A., and Keith, T. (1986). "A simple model of neutrally stratified boundary-layer flow over complex terrain with surface roughness modulations." *Boundary-Layer Meteorology*, 36, 157–186.

Wen, Y.-K., and Chu, S.-L. (1973). "Tornado risks and design wind speed." *J. Struct. Div.*, 99(12), 2409–2421.

Yeatts, B.B., and Mehta, K.C. (1993). "Field study of internal pressures." *Proceedings, 7th U.S. national conference on wind engineering*, UCLA, Los Angeles, 889–897.

Yeatts, B.B., Womble, J.A., Mehta, K.C., and Cermak, J.E. (1994). "Internal pressures for low-rise buildings." *Proceedings, second U.K. conference on wind engineering*, Warwick, England.

Index

A
across-wind response 111, 146
air mean density 146
Analytical Procedure 9–12
    buildings
        commercial with concrete
            masonry unit walls 23–30
        commercial with monoslope
            roof 81–92
        commercial/warehouse 60–77
        non-symmetrical 119–130
        office 35–46
        unusually shaped 119–130
        U-shaped apartment 94–104
    components and cladding 9
    design wind load 9–12
    domed roof building 111–118
    external pressure coefficient (C_p) 10
        for closed building (GC_p) 10
        for closed low-rise building
            (GC_{pf}) 10
    gust effect factor
        for flexible building (G_f) 10
        for rigid building (G) 10
    internal pressure coefficient for
        closed building (GC_{pi}) 10
    main wind force-resisting system 9
    velocity pressure (q) 8
apartment building, U-shaped 93–104
aspect ratio
    h/D. *See* domed roof building
    h/L. *See* roof pressure coefficient
        (C_p)
    L/B. *See* leeward wall pressure
        coefficient (C_p)
    round building 113

B
basic wind speed (V) 25
    Exposure Category 8
    velocity pressure (q) 8
billboard sign 104–111
building types
    apartment 93–104
    commercial 23–34, 59–77, 80–92
    enclosed 143–144
    house, one-story 50–59
    office 35–50
        unusually shaped 119–130
    open 144
        with gable roof 130–134
    partially enclosed 143–144
    storage 130–134
    warehouse 59–77

C

C&C. *See* components and cladding
commercial building
    with concrete masonry unit walls 23–34
        Analytical Procedure 23–30
        Simplified Procedure 31–34
    with monoslope roof 80–92
commercial/warehouse building 59–77
    all height provisions 59–71
    low-rise building provisions 72–77
    wall component pressures 68–71
components and cladding
    Analytical Procedure 9
    buildings
        open 133–134
        round 118
    design force 111
    design wind load
        closed building 11–12
        low-rise building 11
        open building 12
    effective wind area 42, 67, 91, 99, 103, 127
    external pressure coefficient for closed building (GC_p) 10, 13
    gable truss 140
    internal pressure coefficient for closed building (GC_{pi}) 10
    main wind force-resisting system 7, 140
    parapet load 46
    pressure coefficient 143
    roof overhang 91
    roof pressure coefficient (C_p) 46, 103, 142
    roof truss 142
    Simplified Procedure 9
    tilt-up wall system 140
    tributary area 140
construction loads 146
controlling design pressure 46, 103, 104, 118, 128

D

design force 106–111
design wind load
    buildings
        closed 10–12
        flexible 11
        low-rise 11
        open 12
    evaluation methods 7
    other structures 12
    See also Analytical Procedure
    See also design wind pressure
    See also Simplified Procedure
design wind pressure
    controlling 46, 103, 104, 118
    effective wind area 142
    minimum 123, 132
    overhang 91
    roof 91
        joist 92
    trusses 134
    See also design wind load
directional coefficients 4–5, 8, 139
domed roof building 111–118
    roof pressure 114–115
    wall pressure 113–114
    wind load case 116

E

effective wind area 28, 42, 67, 91, 99, 142–143
    components and cladding 103, 127
    external pressure coefficient for closed building (GC_p) 143
    fastener 67
    roof pressure coefficient (C_p) 46
    tributary area 142–143
escarpment effects 49–50
Exposure Category 139–140
    basic wind speed (V) 8
    clearings 139
    topographic factor (K_{zt}) 139–140
exposure velocity pressure coefficient (K_z) 8, 138
external pressure coefficient (C_p) 10
    monoslope roof 82–83
    U-shaped building 95
    See also leeward wall pressure coefficient (C_p)
    See also roof pressure coefficient (C_p)
external pressure coefficient for closed building (GC_p) 10
    effective wind area 42, 67, 91, 99, 143
    equations 13
external pressure coefficient for closed low-rise building (GC_{pf}) 10

F

fastener. *See* wall component pressures
force coefficient (C_f) 144–145
fundamental frequency 107–109, 141

G

gable truss 140
girt. *See* wall component pressures
gust effect factor
    for flexible building (G_f) 4, 10, 11
        fundamental frequency 141
    for rigid building (G) 10, 26, 37, 62, 132
        fundamental frequency 107–109

H

hip roof 104
house with gable/hip roof 50–59

I

impact resistance 145–146
importance factor (I)
    design wind load 9
    Simplified Procedure 9
    velocity pressure (q) 8
interior walls 146
internal pressure (q_i) 10–11
internal pressure coefficient for closed building (GC_{pi}) 10, 26, 144
    interior walls 146
    internal pressure evalutaion 82
    monoslope roof 82
    partially enclosed building 144

L

L/B ratio. *See* leeward wall pressure coefficient (C_p)
leeward wall pressure coefficient (C_p) 38, 62, 82, 95
load combinations 138–139
    allowable stress 138
    trusses 134
    wind directionality factor (K_d) 139
load factor. *See* load combinations

M

main wind force-resisting system
    Analytical Procedure 9
    asymmetrical structure 52–56
    buildings
        open 131–133
        round 113
    components and cladding 7, 140
    definition 7
    design force 106–109
    design wind load
        closed building 10–11
        flexible building 11
        low-rise building 11, 73–74
        open building 12
    domed roof building 113, 114–116
    Exposure Category 139
    external pressure coefficient (C_p) 10
        for closed low-rise building (GC_{pf}) 10
    gable truss 140
    gust effect factor
        for flexible building (G_f) 10
        for rigid building (G) 10
    internal pressure coefficient for closed building (GC_{pi}) 10
    low-rise v. all heights provisions 141
    minimum design wind pressure 123, 132, 145
    net pressure coefficient for open building (C_N) 10
    parapet load 41
    pressure coefficient 141, 143
    roof pressure coefficient (C_p) 27
    roof pressure coefficient for closed low-rise building (GC_{pf}) 75
    Simplified Procedure 2, 9
    tilt-up wall system 140
    tributary area 91, 140
minimum design wind pressure 123, 132, 145

N

net pressure 115–116
net pressure coefficient for open building (C_N) 10

O

office building 35–46
    on escarpment 47–50
open building
    with gable roof 130–134

P

panel. *See* wall component pressures
parapet load
    components and cladding 46
    main wind force-resisting system 41
positive internal pressure evaluation 82
pressure coefficient 141–144
    balcony 141
    components and cladding 143
    main wind force-resisting system 141, 143
    overhang 141
    reduction 29, 127
    roof 142
        shape 143
    See also exposure velocity pressure coefficient (K_z)
    See also external pressure coefficient (C_p)
    See also external pressure coefficient for closed building (GC_p)
    See also external pressure coefficient for closed low-rise building (GC_{pf})
    See also internal pressure coefficient for closed building (GC_{pi})
    See also roof pressure coefficient (C_p)
    See also roof pressure coefficient for closed low-rise building (GC_{pf})
pressure coefficient reduction 29, 127

R

roof overhang 91, 141
roof pressure coefficient (C_p)
    components and cladding 46, 103, 142
    critical loading 39, 121
    effective wind area 46
    load combinations 39, 121
    main wind force-resisting system 27
    monoslope roof 82–83
    roof slope 27
    Simplified Procedure 33
    wind normal to ridge 95
    wind parallel to ridge 83, 96, 121
roof pressure coefficient for closed low-rise building (GC_{pf})
    main wind force-resisting system 75

S

Simplified Procedure 8–9
    commercial building with concrete masonry unit walls 31–34
    components and cladding 9
    design roof pressure 33
    design wind load 9
    domed roof building 111
    importance factor (I) 9
    main wind force-resisting system 9
    multiplying factor (λ) 9
    non-symmetrical building 94, 119
    requirements 32, 81
    roof slope 81
    topographic effects 3
    torsional loading 2
    use 2, 3
Standard
    across-wind response 111, 146
    basic wind speed (V) 25
    changes and additions 2–3
    directional coefficients 4–5
    effective wind area 28
    flexible building 11, 141
    gust effect factor for rigid building (G) 26, 62
    gust effect factors 10
    hip roof 104
    impact resistance 145–146
    internal pressure coefficient for closed building (GC_{pi}) 62
    limitations 3–5
    low-rise building 11, 72
        v. all heights provisions 94, 119
    main wind force-resisting system 7, 123, 132
    minimum design wind pressure 123, 132, 145
    parapet load 41
    pressure coefficient reduction 29
    roof pressure coefficient for closed low-rise building (GC_{pf}) 75
    serviceability states 4
    shape limitations 4–5
    sign convention 10
    Simplified Procedure 9, 32
    technical literature 5
    tributary area 140
    velocity pressure (q) 4
    wind load case 42, 56, 63, 84, 98, 116
strut purlin. *See* wall component pressures

T

terrain exposure. *See* Exposure Category
tilt-up wall system 140
topographic factor (K_{zt}) 8, 139–140
torsional load case. *See* wind load case
tributary area 91, 140, 142–143
truss
    gable 140
    load combinations 134
    roof 142

U

unusually shaped building 119–130

V

velocity pressure (q)
    Analytical Procedure 8
    basic wind speed (V) 8
    comment 145
    equation 8
    escarpment effects 49–50
    Exposure Category 4
    exposure velocity pressure coefficient (K_z) 8
    importance factor (I) 8
    low-rise building provisions 72–73
    q_z v. q_h 145
    reference wind speed 4
    topographic factor (K_{zt}) 8
    wind directionality factor (K_d) 8
vortex shedding 4, 146

W

wall component pressures 68–71
wind directionality factor (K_d) 8, 139
wind load case
    all height provisions 63
    asymmetrical structure 56
    buildings
        low-rise 75–77
        round 116
    diaphragm flexibility 63
    exceptions 123
    requirements 42, 56, 63, 84, 98, 116
    torsional load 77

    torsional wind effects 123
wind load provisions 7–13
    Analytical Procedure 9–12
    design procedure 8–12
    equations for graphs 13
    format 7–8
    Simplified Procedure 8–9
    velocity pressure (q) 8
    Wind Tunnel Procedure 12

wind speed 3, 137–138
    accuracy 137
    exposure velocity pressure coefficient (K_z) 138
    hurricane winds 138
    project location 138
    special wind regions 3, 138
    tornado winds 3
Wind Tunnel Procedure 12

About the Authors

Kishor C. Mehta, Ph.D., P.E., Dist.M.ASCE, is Horn Professor of Civil Engineering and the founder and former director of the Wind Science and Engineering Research Center at Texas Tech University, Lubbock, Texas. He served as Chairman of the ASCE 7 Task Committee on Wind Loads, which produced ASCE 7-88 and ASCE 7-95. He was lead author of the *Guide to the Use of Wind Load Provisions of ASCE 7-95, ASCE 7-98,* and *ASCE 7-02.* Dr. Mehta is past president of the American Association of Wind Engineering and past chairman of the Committee on Natural Disasters, National Research Council. He has been involved in research and education related to wind loads for the past forty years. He directed the ten-year cooperative project between Colorado State University and Texas Tech University that was sponsored by NSF. He also directed the Texas Tech/National Institute of Standards and Technology Cooperative Agreement for Windstorm Damage Mitigation. He was Principal Investigator for the NSF-sponsored Integrative Graduate Education and Research Training program, which established the Ph.D. degree in Wind Science and Engineering at Texas Tech, the only one of its kind in the country. In April 2000, the National Hurricane Conference honored Dr. Mehta with an award for distinguished service in wind engineering, and in 2004 he was elected to membership in the National Academy of Engineering.

William L. Coulbourne, P.E., is the director of wind and flood hazard mitigation for the Applied Technology Council, with his office located in Rehoboth Beach, Delaware. He is a member of the ASCE 7 Wind Load Task Committee and the ASCE 7 Main Committee. He is also a member of the ASCE 7 Flood Load Task Committee and the ASCE 24 Flood Resistant Design and Construction Standard Committee. He has been a practicing structural engineer since 1992 and has been involved in engineering issues related to disasters since 1995. Mr. Coulbourne has participated in mitigation and damage assessment reports after many natural and man-made events for both ASCE and FEMA, including Hurricanes Marilyn, Fran, Bertha, George, Ivan, Charley, and Katrina, the World Trade Center Building Performance Study, flooding in the Red River in 1997, Midwest flooding in 2008, the tornado outbreak in 1999 in Kansas and Oklahoma, and the Greensburg, Kansas, tornado in 2007. He presents workshops and training classes for both ASCE and FEMA on coastal construction practices. These classes have been given in almost 20 states. He has presented papers at ASCE/SEI Structures Congresses, ASFPM conferences, state SEA affiliate meetings, and to other professional organizations whose members are interested in the effects of natural and man-made hazards. Mr. Coulbourne is a licensed Professional Engineer in Delaware, Maryland, and Virginia, and a Certified Structural Engineer.